Nutrigenomics

Carsten Carlberg • Stine Marie Ulven
Ferdinand Molnár

Nutrigenomics

Carsten Carlberg
Institute of Biomedicine
University of Eastern Finland
Kuopio, Finland

Stine Marie Ulven
Department of Nutrition
University of Oslo
Oslo, Norway

Ferdinand Molnár
School of Pharmacy
University of Easterm Finland
Kuopio, Finland

ISBN 978-3-319-80803-1 ISBN 978-3-319-30415-1 (eBook)
DOI 10.1007/978-3-319-30415-1

Softcover reprint of the hardcover 1st edition 2016

Printed on acid-free paper

This Springer imprint is published by Springer Nature
The registered company is Springer International Publishing AG Switzerland

Preface

Our daily diet is more than a collection of carbohydrates, lipids and proteins that provide energy and serve as building blocks of our life; our diet is also the most dominant environmental signal to which we are exposed from womb to death. The fascinating area of *nutrigenomics* analyses this daily communication between diet, food and nutrients, their metabolites and our genome. This book describes how nutrition shapes human evolution and demonstrates its consequences for our susceptibility to diseases, such as diabetes and atherosclerosis. Inappropriate diet can yield stress for our cells, tissues and organs and then it is often associated with low-grade chronic inflammation. Overnutrition paired with physical inactivity leads to overweight and obesity and results in increased burden for a body that originally was adapted for a life in the savannas of East Africa. Therefore, this textbook does not discuss a theoretical topic in science, but it talks about real life and our life-long "chat" with diet. We are all food consumers, thus each of us is concerned by the topic of this book and should be aware of its mechanisms.

The availability of the sequence of the complete human genome and the consequent development of next-generation sequencing technologies have significantly affected nearly all areas of bioscience. This new knowledge was the starting point for new disciplines, such as genomics and its sub-discipline nutrigenomics. The latter comprises not only the variation of the human genome, such as single nucleotide polymorphisms (SNPs), but also the dynamic packaging of the genome into chromatin including all information stored in this epigenome. Moreover, this book describes the proteins that are involved in the signal transduction between dietary molecules and the genome, such as nuclear receptors, chromatin modifiers and energy status-sensing kinases, and their mechanism of action.

The purpose of this book is to provide an overview on the principles of nutrigenomics and their relation to health or disease. We are not aiming to compete with more comprehensive textbooks on molecular nutrition, evolutionary biology, genomics, gene regulation or metabolic diseases, but rather will focus on the essentials and will combine, in a compact form, elements from different disciplines. In order to facilitate the latter, we favour a high figure-to-text ratio following the rule "a picture tells more than thousand words".

The content of this book is based on the lecture course "Nutrigenomics", which is held since 2003 once per year by one of us (C. Carlberg) at the University of Eastern Finland in Kuopio. The book is subdivided into 3 sections and 12 chapters. Following the "Introduction", there are sections on the "Molecular genetic basis" and the "Links to disease", which take a view on nutrigenomics from the perspective of molecular mechanisms or from the causes of metabolic diseases, respectively.

This book is primarily designed for master level students of biosciences, but may also be used by students of other biomedical disciplines and by PhD students. The book has five major learning objectives. Students should:

(i) Get an overview on human variation on the level of the genome and epigenome, in response to dietary molecules and the regulatory proteins involved in the respective signal transduction processes
(ii) Have an understanding of the diseases that are strongly associated with dietary intake and physical inactivity, such as obesity, type 2 diabetes, atherosclerosis and the metabolic syndrome
(iii) Recognize the key components and mechanisms in nutrigenomics and the multiple layers of its regulatory complexity
(iv) Show the ability to analyze the human genome and epigenome and its variation in nutrition sensing and information processing processes, i.e. to judge their impact for on the complex etiology of metabolic diseases
(v) Apply knowledge in nutrigenomics in designing, performing and analyzing respective experiments, such as quantitative PCR, RNA-seq or ChIP-seq

We hope the readers will enjoy this demonstrative book and get as enthusiastic about the topic of nutrigenomics as the authors do.

Kuopio and Oslo
December 2015

Carsten Carlberg
Stine Marie Ulven
Ferdinand Molnár

Acknowledgements

The authors would like to thank Reinhard Bornemann, MD, PhD, Marjukka Kolehmainen, PhD, Vibeke Telle-Hansen, PhD, and Jenni Puurunen, MSc, for extensive proofreading and constructive criticism.

Contents

Abbreviations

$1,25(OH)_2D_3$	1,25-dihydroxyvitamin D_3
$25(OH)D_3$	25-hydroxyvitamin D_3
5,10-MTHF	5,10-methylene THF
α-MSH	α-melanocyte-stimulating hormone
ABC	ATP-binding cassette
ABL	abetalipoproteinemia
AC	adenylate cyclase
ACAT1	acetyl-CoA acetyltransferase 1
ACC	acetyl-CoA carboxylase
ACL	ATP citrate lyase
ADRB3	adrenoceptor beta 3
AGRP	agouti-related peptide
AKT	Akt murine thymoma viral oncogene homolog
ALA	α-linolenic acid
ALOX5	arachidonate 5-lipoxygenase
ALOX15	arachidonate 15-lipoxygenase
AMPK	adenosine monophosphate-activated protein kinase
AMY1	amylase
ANGPTL2	angiopoietin-like protein 2
APEH	N-acylaminoacyl-peptide hydrolase
AP-1	activating protein 1
APO	apolipoprotein
AR	androgen receptor
ARL4C	ADP-ribosylation factor-like
ARNTL	aryl hydrocarbon receptor nuclear translocator-like
ASC	apoptosis-associated speck
ASIP	agouti signaling protein
ATF6	activating transcription factor 6
ATM	ataxia telangiectasia mutated
atRA	all-*trans* retinoic acid
β-OHB	β-hydroxybutyrate

BAAT	bile acid-CoA-amino acid N-acetyltransferase
BAT	brown adipose tissue
BDNF	brain-derived neurotrophic factor
BLK	B lymphoid tyrosine kinase
BMD	bone mineral density
BMI	body mass index
BMP	bone morphogenetic protein
bp	base pair
CAMKK	Ca^{2+}/calmodulin-dependent protein kinase kinase
CAMP	cathelicidin anti-microbial peptide
CAR	constitutively androstane receptor
CASP	caspase
CBL	Cbl proto-oncogene, E3 ubiquitin protein ligase
CBP	CREB-binding protein
CCL	chemokine (C-C motif) ligand
CCR	C-C chemokine receptor
CD36	CD36 molecule
CDC42	cell-division cycle 42
CDKAL1	CDK5 regulatory subunit associated protein 1-like 1
CDKN	cyclin-dependent kinase inhibitor
CDP	common dendritic cell progenitor
CEBP	CCAAT-binding protein
CEL	carboxyl ester lipase
CETP	cholesterol ester transfer protein
CHD	coronary heart disease
ChIP	chromatin immunoprecipitation
CITED1	CBP/p300-interacting transactivator 1
CLOCK	clock circadian regulator
CNR	cannabinoid receptor 1
CNS	central nervous system
CNV	copy number variant
CPT1A	carnitine palmitoyltransferase 1A
CREB3L3	cAMP responsive element binding protein 3-like 3
CRP	C-reactive protein
CRTC2	CREB-regulated transcription co-activator 2
CRY1	cryptochrome circadian clock 1
CSF2	colony stimulating factor 2
CVD	cardiovascular disease
CXCL5	chemokine (C-X-C motif) ligand 5
CXCR2	CXC-chemokine receptor 2
CYP	cytochrome P450
D2HGDH	D-2-hydroxyglutarate dehydrogenase
DAG	diacylglycerol
DALY	disability-adjusted life-year
DAMP	damage-associated molecular pattern

DGAT1	diacylglycerol O-acyltransferase 1
DHA	docosahexaenoic acid
DHF	dihydrofolate
DEFB4	defensin, beta 4A
DNMT	DNA methyltransferase
EHMT1	euchromatic histone-lysine N-methyltransferase 1
EGIR	European Group for the study of Insulin Resistance
EIF2A	eukaryotic translation initiation factor 2A
EIF2AK3	eukaryotic translation initiation factor 2-alpha kinase 3
ENCODE	encyclopedia of DNA elements
ENPP1	ectonucleotide pyrophosphatase/phosphodiesterase 1
EPA	eicosapentaenoic acid
EPIC	European Prospective Investigation into Cancer and Nutrition
eQTL	expression quantitative trait locus
ER	endoplasmatic reticulum
ERN1	endoplasmic reticulum to nucleus signaling 1
ES cell	embryonic stem cell
E%	percent of total energy
FABP6	ileal fatty acid-binding protein 6
FAD	flavin adenine dinucleotide
FAIRE	formaldehyde-assisted isolation of regulatory elements
FAO	Food and Agriculture Organization
FAS	Fas cell surface death receptor
FASN	fatty acid synthase
FFA	free fatty acid
FGF	fibroblast growth factor
FGFR4	FGF receptor 4
FH	fumarate hydratase
FOX	forkhead box
FTO	fat mass and obesity associated
FXR	farnesoid X receptor
G6PC	glucose-6-phosphatase
GAB1	GRB2-associated binder 1
GCK	glucokinase
GC-MS	gas chromatography-mass spectrometry
GDH	glutamate dehydrogenase
GH	growth hormone
GLP1	glucagon-like peptide 1
GO	gene ontology
GPAT	glycerol phosphate acyl transferase
GPR	G-protein-coupled receptor
GR	glucocorticoid receptor
GRB	growth factor receptor-bound protein
GS	glycogen synthase
GSK3	glycogen synthesis kinase 3

GWAS	genome-wide association study
HAT	histone acetyltransferase
HDAC	histone deacetylase
HDM	histone demethylase
HDL	high-density lipoprotein
HHEX	hematopoietically expressed homeobox
HIF1α	hypoxia-inducible factor 1α
HIV	human immunodeficiency virus
HLA	human leukocyte antigen
HMGCR	3-hydroxy-3-methylglutaryl-CoA reductase
HMGCS2	3-hydroxy-3-methylglutaryl-CoA synthase 2
HMT	histone methyltransferase
HNF	hepatocyte nuclear factor
HPT	hypothalamic-pituitary-thyroid
HSF1	shock transcription factor 1
HSP	heat-shock protein
HTG	hypertriglycerolemia
IAP	intracisternal A particle
ICAM1	intercellular adhesion molecule 1
IDF	International Diabetes Federation
IDH	isocitrate dehydrogenase
IDOL	inducible degrader of LDLR
IFG	impaired fasting glucose
IFNγ	interferon gamma
IGF	insulin-like growth factor
IGF1R	IGF1 receptor
IGF2BP2	insulin-like growth factor 2 mRNA binding protein 2
IGT	impaired glucose tolerance
IKBK	inhibitor of kappa light polypeptide gene enhancer in B cells, kinase
IL	interleukin
IL1R	IL1 receptor
IL1RN	IL1 receptor antagonist
indel	insertion-deletion variant
IR	insulin receptor
IRF3	interferon-regulatory factor 3
IRS	insulin receptor substrate
IRX3	iroquois homeobox 3
ITGA4	integrin, alpha 4
ITGB2	integrin, beta 2
IVF	*in vitro* fertilization
JAK	Janus kinase
K^{ATP}	ATP-sensitive K^+
kb	kilobase
KEAP1	kelch-like ECH-associated protein 1
KLF	Krüppel-like factor

KCNJ11	potassium inwardly rectifying channel, subfamily J, member 11
LA	linolenic acid
LBD	ligand-binding domain
LCAD	long-chain acyl-CoA dehydrogenase
LCAT	lecithin cholesterol acyltransferase
LCT	lactase
LDL	low-density lipoprotein
LDLR	LDL receptor
LDLRAP1	LDLR accessory protein 1
LEP	leptin
LEPR	leptin receptor
LIPC	hepatic lipase
LIPE	hormone sensitive lipase
LIPG	endothelial lipase
LINE	long interspersed element
LPCAT3	lysophospholipid acyltransferase 3
LPL	lipoprotein lipase
LRH-1	liver receptor homolog 1
LRP1	LDLR-related protein 1
LXR	liver X receptor
MAF	minor allele frequency
MAFA	v-maf avian musculoaponeurotic fibrosarcoma oncogene homolog A
MAN2A1	mannosidase, alpha, class 2A, member 1
MAP	mitogen-activated protein
MAPK8	mitogen-activated protein kinase 8 (also called JNK)
MC4R	melanocortin 4 receptor
M-CFU	myeloid stem cells
MCM6	minichromosome maintenance type 6
MDH	malate dehydrogenase
MDP	macrophage and dendritic cell progenitor
MECP2	methyl-CpG-binding protein 2
MED	mediator
MHC	major histocompatibility complex
miRNA	micro RNA
MLXIPL	MLX interacting protein-like
mmHg	millimeters of mercury
MODY	maturity onset diabetes of the young
MPO	myeloperoxidase
MR	mineralocorticoid receptor
mRNA	messenger RNA
MSR1	macrophage scavenger receptor 1
MTHFR	methylenetetrahydrofolate reductase
MTNR1B	melatonin receptor 1B
mTORC	mammalian target of rapamycin complex
MTTP	microsomal triglycerole transfer protein

MYCL	v-myc avian myelocytomatosis viral oncogene lung carcinoma derived homolog
MYD88	myeloid differentiation primary response protein 88
MYF5	myogenic factor 5
NAD	nicotinamide adenine dinucleotide
NAFLD	non-alcoholic fatty liver disease
NAMPT	nicotinamide mononucleotide phosphoribosyltransferase (also called visfatin)
NANOG	NANOG homeobox
NCEH1	neutral cholesterol ester hydrolase 1
NCOA	nuclear receptor coactivator
NCEP	National Cholesterol Education Program
ncRNA	non-coding RNA
NEUROD1	neuronal differentiation 1
NF-κB	nuclear factor κB
NLRP	NLR protein
NLR	NOD-like receptor
NO	nitric oxide
NOS2	inducible nitric oxide synthase 2
NPC1L1	Niemann-Pick C1-like protein 1
NPY	neuropeptide Y
NRIP1	nuclear receptor interacting protein 1
NSAID	non-steroidal anti-inflammatory drug
NTS	nucleus tractor solitarius
OAADR	O-acetyl ADP-ribose
O-GlcNAc	O-linked N-acetylglucosamine
OGA	O-GlcNAcase
OGT	O-GlcNAc transferase
OGTT	oral glucose tolerance test
PAMP	pathogen-associated molecular pattern
PAX	paired box
PC	pyruvate carboxylase
PCK	phosphoenolpyruvate carboxykinase
PCSK1	POMC proprotein convertase subtilisin/kexin type 1
PDH	pyruvate dehydrogenase
PDPK	3-phosphoinositide dependent protein kinase
PDX1	pancreatic and duodenal homeobox 1
PER1	period circadian clock 1
PFKFB2	6-phosphofructo-2-kinase/fructose-2,6-biphosphatase 2
PGE2	prostaglandin E2
PH	pleckstrin-homology
PI3K	phosphoinositide 3-kinase
PIP3	phosphatidylinositol-3,4,5-triphosphate
PKA	protein kinase A
PLAU	plasminogen activator, urokinase

PLTP	phospholipid transfer protein
PNPLA	patatin-like phospholipase domain containing
Pol II	RNA polymerase II
POMC	pro-opiomelanocortin
POU5F1	POU class 5 homeobox 1
PPAR	peroxisome proliferator-activated receptor
PPARGC1A	PPAR gamma, coactivator 1 alpha
PPP2	protein phosphatase 2
PRDM16	PR domain containing 16
PRR	pattern recognition receptor
PUFA	polyunsaturated fatty acid
PTB	phosphotyrosine-binding
PTEN	phosphatase and tensin homologue
PTGS2	prostaglandin-endoperoxide synthase 2 (also known as COX2)
PTPN1	protein tyrosine phosphatase, non-receptor type 1
PVN	paraventricular nuclei
PXR	pregnane X receptor
qPCR	quantitative PCR
RAPTOR	regulatory associated protein of TOR
RAR	retinoic acid receptor
RBP4	retinol binding protein 4
RE	response element
REV-ERB	Reverse-Erb
RHEB	Ras homolog enriched in brain
RHOQ	Ras homolog family, member Q
RIG1	retinoic acid-inducible gene 1
RLR	RIG1-like helicase receptors
RNAi	RNA interference
ROR	RAR-related orphan receptor
ROS	reactive oxygen species
RRAG	Ras-related GTP binding
RPS6K	ribosomal protein S6 kinase
RXR	retinoid X receptor
S6K	S6 kinase
SAH	S-adenosylhomocysteine
SAM	S-adenosylmethionine
SCAP	SREBF chaperone
SCD1	steroyl-CoA desaturase 1
SCN	suprachiasmatic nucleus
SCNN1	sodium channel, non-voltage-gated 1
SDH	succinate dehydrogenase
SEC16B	SEC16 homolog B
SERPINE1	serpin peptidase inhibitor, clade E (also called PAI-1)
SF-1	steroidogenic factor 1
SFA	saturated fatty acids

SFRP5	frizzled-related protein 5
SHC	Src homology 2 domain containing
SI	sucrase-isomaltase
SIM1	single-minded family bHLH transcription factor 1
SINE	short interspersed element
siRNA	small interfering RNA
Sir2	silent information regulator 2
SIRT	sirtuin
SLC	solute carrier
SLCO	solute organic anion transporter
SLK	STE20-like kinase
SNP	single nucleotide polymorphism
SNS	sympathetic nervous system
SOCS3	suppressor of cytokine signaling 3
SOD2	superoxide dismutase 2
SORBS1	sorbin and SH3 domain containing 1
SOS	son of sevenless
SPI1	spleen focus forming virus proviral integration oncogene (also called PU.1)
SREBF1	sterol regulatory element-binding transcription factor 1
STAT	signal transducer and activator of transcription
SULT2A1	sulfotransferase family 2A, member 1
T1D	type 1 diabetes
T2D	type 2 diabetes
TAS1R2	taste receptor, type 1, member 2
TBC1D	TBC1 domain family, member 1
TBP	TATA-box binding protein
TCA	tricarboxylic acid
TD	Tangier disease
TET	ten-eleven translocation
TGFB1	transforming growth factor beta 1
T_H	T helper
THF	tetrahydrofolate
THRSP	thyroid hormone responsive
TIFIA	transcription initiation factor IA
TLR	Toll-like receptor
TMEM18	transmembrane protein 18
TNF	tumor necrosis factor
TNFR	TNF receptor
TOR	target of rapamycin
TRAF2	TNF receptor-associated factor 2
T_{REG}	regulatory T
TSC2	tuberous sclerosis 2
TSS	transcription start site
UBR1	ubiquitin protein ligase E3 component n-recognin 1

UDP	uridine diphosphate
UCP	uncoupling protein
UGT2B4	UDP glucuronosyltransferase 2 family, polypeptide B4
UNC5B	unc-5 homolog B
UTR	untranslated region
UV	ultraviolet
VCAM1	vascular cell adhesion molecule 1
VDR	vitamin D receptor
VLDL	very low-density lipoprotein
VNN1	vanin 1
WAT	white adipose tissue
WHO	World Health Organization
WHR	waist-hip ratio
YWHA	tyrosine 3-monooxygenase/tryptophan 5-monooxygenase activation protein (also called 14-3-3)
XBP1	X-box binding protein 1

Part I
Introduction

Chapter 1
Nutrition and Common Diseases

Abstract A significant lifestyle change has happened during the last 100–200 years with industrialization, rapid urbanization, economic development and market globalization. Changes in food intake and a more sedentary lifestyle both increase the risk of chronic non-communicable diseases, such as obesity, type 2 diabetes (T2D), cardiovascular disease (CVD) and various cancers. Diet is one of the key environmental factors particularly involved in the pathogenesis and progression of most of these diseases. Together with physical inactivity, alcohol abuse and tobacco use, these four key environmental factors cause metabolic and physiological changes, such as overweight and obesity (Chap. 8), insulin resistance and β cell failure (Chap. 9), T2D (Chap. 10), hypertension (Chap. 11), dyslipidemia leading to atherosclerosis and cardiovascular failure (Chap. 11) and the metabolic syndrome (Chap. 12). However, not a single individual food component but the interaction between many of them and the overall quality of diet is responsible for the increased risk for these diseases.

In this chapter, we will provide a first overview of the role of nutrition in health and disease. We will describe the evidence of dietary factors in non-communicable diseases and the impact of exercise on the prevention of diseases. Moreover, we will describe low-grade chronic inflammation (Chap. 7) as the underlying cause of many non-communicable diseases. We will use obesity and cancer as examples, in order to describe the link between inflammation and nutrition-triggered diseases.

Keywords Nutrition • Non-communicable diseases • Cancer • Obesity • Physical activity • Adipose tissue • Inflammation

1.1 Human Nutrition

Diet is composed of food groups that collectively provide the human body with its nutritional needs of macro- and micronutrients. In addition to nutrients, food also contains hundreds of bioactive compounds that have an effect on metabolism but their functions on human health are more or less unknown. The main food groups in our diet includes cereals and cereals products, fruit and vegetables, milk and dairy products, meat, fish, and fats and oils. Some of the challenges in studying the role of single nutrients or food groups on human health are the relatively small effects of

C. Carlberg et al., *Nutrigenomics*, DOI 10.1007/978-3-319-30415-1_1

each food item or nutrient. It is the overall quality of diet and the interaction among many food groups and nutrients, and not an individual food component that is playing a role in human health. Therefore, increased emphasis in nutrition research has been placed on understanding healthy dietary patterns, such as the Mediterranean diet and the Nordic diet, on human health.

Human nutrition is the provision to obtain a sufficient amount of macro- and micronutrients necessary to support essential functions of life, such as energy supply, reproduction and growth, and to maintain good health. The main macronutrients in diet are carbohydrates (digested to glucose), proteins (digested to amino acids) and lipids (dissoluted to fatty acids and cholesterol), whereas micronutrients mainly consist of vitamins and minerals (Chap. 3). Macronutrients provide the body with energy when they are metabolized or are stored in form of glycogen, proteins or triacylglycerols when there is excess of energy. Equally, these nutritional compounds can be used in anabolism for the synthesis of new macronutrients. Both catabolism and anabolism are tightly controlled pathways composed of enzymatic cascades (Fig. 1.1).

The key metabolic tissues are intestine, liver, adipose tissue, skeletal muscle and the brain (Chaps. 8, 9, 10, 11, and 12). Food intake relies on the interaction between

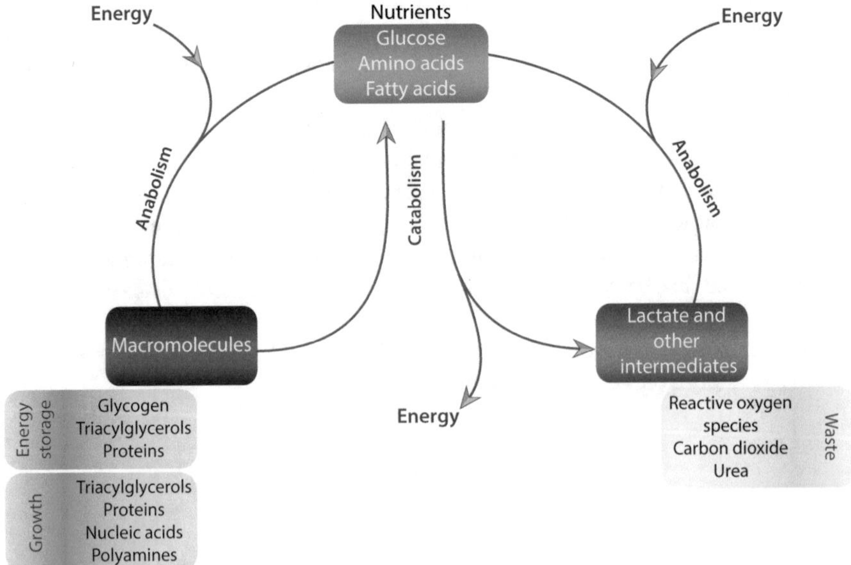

Fig. 1.1 Simplified view of metabolism principles. Macronutrients provide the body with energy via catabolic processes or energy storage of macromolecules, such as glycogen (from glucose), proteins (from amino acids) and triacylglycerols (from fatty acids). When macronutrients are catabolized to generate energy, lactate and other intermediates, such as reactive oxygen species (ROS), carbon dioxide and ammonia, are synthesized. The body needs to get rid of this waste via endogenous anti-oxidant defense systems, respiration via the lungs and urea-excretion via the urine, in order to avoid intoxication. Furthermore, the human body also uses amino acids for the synthesis of nucleic acids and polyamines for normal growth and regeneration

homeostatic regulation and hedonic sensations. During a meal a variety of hormones and nutrients are released by the gastrointestinal tract and associated glands, and these are transported in the blood to the central nervous system (CNS). The palatability (meaning: its hedonic reward) of food is therefore important in regulating its intake and metabolism, in order to match the nutritional need of the organism. After a meal, the liver is the first organ to receive enterally absorbed nutrients and pancreatic hormones via the portal vein. Thus, the liver plays a major role in energy storage, in particular of carbohydrates (Chap. 9), but this organ also metabolizes amino acids and fatty acids. Adipose tissue (Chap. 8) stores most of the body's energy as triacylglycerols, and releases fatty acids from adipose tissue when other tissues need energy in a process called lipolysis. Skeletal muscle is a highly active organ that uses both types of stored energy, i.e. glycogen and triacylglycerols, and takes up glucose and fatty acids directly from the circulation depending on the type of exercise performed (Sect. 9.1). The brain controls satiety and hunger, which makes this organ particularly important for regulating energy balance of the whole body (Sect. 8.4).

Acceptable macronutrient distribution ranges for protein, carbohydrate and fat have been set by considering epidemiological and experimental evidences. Daily consumption of macronutrients within these ranges should maintain health by providing the body with essential nutrients and prevent non-communicable diseases. Dietary carbohydrates consist of polysaccharides, oligosaccharides, disaccharides, such as sucrose, maltose or lactose, and free monosaccharides, such as glucose or fructose. To maintain health and prevent non-communicable diseases, the acceptable distributation range of carbohydrates should represent 45–65 % of total energy (E%) and free sugars should not exceed 10 E%.

Glucose is the most common monosaccharide in the diet and is absorbed directly into the bloodstream during dietary digestion. Since the CNS is largely reliant on glucose as its metabolic fuel, and red blood cells are even entirely dependent on glucose, the plasma concentration of glucose is strictly regulated by the peptide hormones insulin and glucagon (Sect. 9.1). After digestion, glucose is being taken up in all tissues by specific transport proteins and phosphorylated to glucose-6-phosphate (Sect. 3.1) before entering the pathways of glycogen synthesis (storage) or glycolysis (anaerobic energy production). The end product of glycolysis is pyruvate that enters the tricarboxylic acid (TCA) cycle, in order to generate energy more efficiently due to use of oxygen (oxidative phosporylation). During overnight fasting, glycogen is broken down to glucose via glycogenolysis, and the body starts to synthesize glucose from amino acids or other small molecules via gluconeogenesis, in order to ensure energy supply from sufficient concentration of glucose in the blood (5 mM). Nutrients or their metabolites regulate the expression of genes involved in these metabolic pathways either directly (Sect. 6.2) or indirectly via insulin. Therefore, disturbances in insulin signaling (Sect. 9.2) are the major cause of metabolic diseases (Chap. 12).

Proteins are polymers of amino acids. All processes in life, ranging from control of metabolism, via immune functions to physical movement, are dependent on these "workers" of the cell. In humans, 20 amino acids are used as building blocks for proteins. Nine of these amino acids are "essential", i.e. the human body cannot synthesize them itself and thus must obtain them from diet. The acceptable daily intake range of protein is 10–35 E%. In addition, a number of amino acids are used for other purposes than protein synthesis, such as the synthesis of purines and pyrimidines as nucleic acid components. Normally, the rate of amino acid breakdown balances the rate of their intake, i.e. the body does not store amino acids as an energy source. Amino acid metabolism takes place primarily in the liver and the exception is the catabolism of branched chain amino acids in the skeletal muscle. During extreme situations, such as starvation or disease, the body breaks down proteins to amino acids, in order to use their carbon backbone as substrates for the synthesis of glucose, fatty acids and ketone bodies. Thus, amino acids can play an important role in whole body energy homeostasis.

Lipids are the major source of energy for the human body with an acceptable daily intake range of 20–35 E%. Triacylglycerols, i.e. glycerol esterified with three fatty acids, represent most of the dietary fat, while cholesterol and phospholipids are of lower amount. Fatty acids are classified as saturated (SFAs), monounsaturated or polyunsaturated depending on the number of double bonds in their backbone structure. The proportion of the fatty acids in food depends on its source, such as animal or plant origin. The challenge with fat intake is the proportion between saturated and unsaturated fat. The recommended daily intake of monounsaturated fatty acids should be in the order of 10–20 E%, that of SFAs less than 10 E% and that of polyunsaturated fatty acids (PUFAs) 5–10 E% including at least 1 E% as ω-3 fatty acids. Lipids are insoluble in aqueous solutions and are therefore found in all cells as membrane-associated lipids, in adipocytes as droplets of triacylglycerols and in blood plasma as major components of differently sized lipoproteins (Sect. 11.3). Dietary lipids are not only important suppliers of energy (via β-oxidation) but some of them, such as fatty acids, steroid hormones and eicosanoids, also act as co-enzymes and biological active molecules (Box 1.1) and play critical roles in the control of homeostasis (Chaps. 3 and 9). Dyslipidemias, i.e. disturbances in lipid metabolism, are therefore common in chronic metabolic diseases (Sect. 11.4).

Box 1.1 Dietary Components Acting as Signaling Molecules

Biological active molecules either carry signals over long distances (endocrine signaling), act locally to communicate information between neighboring cells (paracrine signaling) or are synthesized in the cell to communicate within the cell itself (autocrine signaling). The lipophilic fraction of these molecules, such as steroid hormones or prostaglandins, can cross the plasma membrane and bind to transcription factors in the cytoplasm or nucleus (Sect. 3.2), whereas the larger hydrophilic fraction binds to membrane proteins on the surface of target cells. Macronutrients, such as fatty acids, cholesterol, glucose and amino acids, and micronutrients, such as vitamin A, vitamin D, vitamin E, calcium and iron, can either act as ligands of nuclear receptors or as co-factors to enzymes. Nutrients can also bind to membrane proteins and initiate intercellular signaling pathways leading to changes in the activity of transcription factors. Target genes regulated by transcription factors are encoding proteins that play important roles in transport, uptake and storage of nutrients and as enzymes in metabolic pathways. Thus, these dietary components play critical roles in the control of energy homeostasis (Chap. 3).

In a historical perspective, the medical focus on diet was to avoid nutrient deficiency diseases, such as scurvy and rickets. In contrast, today the main focus of nutrition research and the definition of nutritional requirements are from the perspective of preventing non-communicable diseases, i.e. the focus is now on dietary patterns that maintain health rather than just on the intake of nutrients (Box 1.2). A joint *World Health Organization* (*WHO*)/*United Nations Food and Agriculture Organization* (*FAO*) report summarizes evidences from nutritional epidemiology and experimental research related to diet for the prevention of non-communicable diseases (see examples in Tables 1.1, 1.2, 1.3 and 1.4). In this context, the term "convincing evidence" is used when large numbers of epidemiological studies show consistent associations between exposure and disease, with little or no evidence to the contrary. These associations are also biological relevant. These evidences have been translated into food-based nutritional guidelines facilitating the understanding what a healty diet consist of, in order to prevent non-communicable diseases and maintain health. These food-based nutritional guidelines focus on food groups rather than nutrient intake and macronutrient distribution ranges for protein, carbohydrate and fat. A healthy diet includes (i) a nutritious diet based on a variety of foods originating mainly from plants rather than from animals, (ii) to eat bread, grains, pasta, rice or potatoes several times a day, (iii) to eat a variety of vegetables and fruits, preferably fresh and local, several times per day, (iv) to control fat intake and replace most SFAs with unsaturated vegetable oils or soft margarines, (v) to replace fatty meat and meat products with beans, legumes, lentils, fish, poultry or lean meat, (vi) to use milk and dairy products that are low in both fat and salt and (vii) to select foods that are low in sugar and eat refined sugar sparingly, limiting the frequency of sugary drinks and sweets. People should also choose a low-salt diet. Total salt intake should not be more than one teaspoon (6 g) per day, including the salt in bread and processed, cured and preserved foods, in order to prevent hypertension (Sect. 11.1).

Box 1.2 Global Burden of Disease

Non-communicable diseases are non-infectious respectively non-transmissible to people, and many of them are chronic diseases with slow progression. They include autoimmune diseases, CVDs, many form of cancer, asthma, T2D, chronic kidney disease, osteoporosis, Alzheimer's disease and more. Among the 56 million deaths occurred in the world in 2012, 38 million (68 %) were due to non-communicable diseases, with CVD, cancer, T2D and chronic respiratory diseases being the main causes. According to the *WHO*, the number of total deaths caused by non-communicable diseases, in particular cancer, heart disease and stroke, will increase until 2030 by 37 % to 52 million yearly deaths globally. The greatest increases will be in Africa, South-East Asia and the Eastern Mediterranean.

1.2 Nutrition and Obesity

In 2010, overweight and obesity were estimated to cause 3.4 million deaths and 3.9 % of years of life lost worldwide. The global prevalence of obesity (defined via a body mass index (BMI) ≥ 30, which is calculated as weight in kilograms divided by the square of the height in meters) doubled between 1980 and 2014. In 2014, the worldwide number of obese adults (BMI ≥ 30) was 600 millions, but the total number of overweight adults (BMI ≥ 25) was 1.9 billion (Sect. 8.1). In 2013, the worldwide number of overweight children under the age of 5 years was 42 million. The United States had the largest absolute increase in the number of obese people since 1980, followed by China, Brazil and Mexico. The age-standardized mean BMI ranges from less than 22 in parts of sub-Sahara Africa and Asia to 30 to 35 in some Pacific islands and countries in the Middle East and North Africa. The risk of T2D (Sect. 10.1) and hypertension (Sect. 11.1) rises with increased weight, while there is overlap between the prevention of obesity and the prevention of a variety of chronic diseases, especially T2D.

A dietary factor that convincingly reduces the risk of weight gain and obesity is dietary fibre (Table 1.1). Regular physical activity also convincingly reduces the

Table 1.1 Overview of lifestyle factors and risk of developing obesity

Evidence	Decreased risk	No relationship	Increased risk
Convincing	Regular physical activity		Sedentary lifestyles
	High dietary intake of non-starch polysaccharides (dietary fibre)[a]		High intake of energy-dense micronutrient-poor foods[b]
Probable	Home and school environments that support healthy food choices for children[c]		Heavy marketing of energy-dense foods[c] and fast-food outlets[c]
			High intake of sugars-sweetened soft drinks and fruit juices
	Breastfeeding		Adverse socioeconomic conditions[c] (in developed countries, especially for women)
Possible	Low glycemic index foods	Protein content of the diet	Large portion sizes
			High proportion of food prepared outside the home (developed countries)
			"Rigid restraint/periodic disinhibition" eating patterns
Insufficient	Increased eating frequency		Alcohol

[a]Specific amounts will depend on the analytical methodologies used to measure fibre
[b]Energy-dense and micronutrient-poor foods tend to be processed foods that are high in fat and/or sugars. Low energy-dense (or energy-dilute) foods, such as fruit, legumes, vegetables and whole grain cereals, are high in dietary fibre and water
[c]Associated evidence and expert opinion included

risk (Sect. 1.6). In contrast, the dietary factor that convincingly increases the risk of weight gain and obesity is high intake of energy-dense foods that are not only highly processed but also micronutrient poor. Typical energy-dense foods are high in fat (butter, oils and fried foods), sugar or starch. In contrast, energy-dilute foods have high content of water and fibre, such as fruits, vegetables, legumes and whole grain cereals. Other dietary factors that have shown a probably increased risk of weight gain and obesity are high intake of sugar-sweetened soft drinks and fruit juices. Thus, in order to prevent obesity people should limit energy intake from total fats and sugars, and increase consumption of fruit and vegetables, as well as legumes, whole grains and nuts. People should also be engage in regular physical activity.

1.3 Nutrition and Cancer

In 2012 the worldwide estimated number of persons with recent diagnosis of cancer was 14.1 millions. This number is expected to increase to some 24 millions by 2035. In 2012, cancer caused wordwide 8.2 million deaths (representing 22 % of non-communicable disease deaths). Cancer is defined as a disease, in which the normal control of cell division is lost, so that an individual cell multiplies inappropriately to form a primary tumor. The tumor cells may eventually spread through the body and form metastases, both primary and secondary cancer finally causing suffering and death. Cancer can arise from different tissues and organs in the body, thus there are many different forms of cancer.

Common variations in genes that regulate the metabolism of dietary molecules on their own do not cause any significant cancer risk, but in combination with an unhealthy lifestyle, such as smoking and high-fat diet, they may do so. Thus, dietary molecules can both increase or reduce the cancer risk, but compared with smoking their effects are usually small.

Also according to the *WHO*, diet is one of the most important environmental factors contributing to cancer risk. Around one third of cancer deaths are due to high body mass index, low fruit and vegetable intake, lack of physical activity, tobacco use and alcohol use. Even if dietary factors account for approximately 30 % of cancer cases in industrialized countries, only a few definite relationships between specific nutrient-related factors and cancer are established.

For example, there is convincing evidence that overweight and obese individuals have increased risk of cancer of esophagus, colorectum, breast in post-menopausal women, endometrium and kidney, while individuals who consume a high amount of alcohol are prone to cancers of the oral cavity, pharynx, larynx, esophagus, liver and breast. Moreover, aflatoxins and some forms of salted and fermented fish contribute to the development of liver and nasopharynx cancer, respectively (Table 1.2).

Table 1.2 Overview of lifestyle factors and risk of developing cancer

Evidence	Decreased risk	Increased risk
Convincing[a]	Physical activity (colon)	Overweight and obesity (esophagus, colorectum, breast in postmenopausal women, endometrium, kidney)
		Alcohol (oral cavity, pharynx, larynx, esophagus, liver, breast)
		Aflatoxin (liver)
		Chinese-style salted fish (nasopharynx)
Probable[a]	Fruits and vegetables (oral cavity, esophagus, stomach, colorectum[b])	Preserved meat (colorectum)
	Physical activity (breast)	Salt-preserved foods and salt (stomach)
Possible/insufficient	Fibre	Very hot (thermally) drinks and food (oral cavity, pharynx, esophagus)
		Animal fats
		Heterocyclic amines
	Soya	Polycyclic aromatic hydrocarbons
	Fish	Nitrosamines
	ω-3 Fatty acids	
	Carotenoids	
	Vitamins B_2, B_6, folate, B_{12}, C, D, E	
	Calcium, zinc and selenium	
	Non-nutrient plant constituents (e.g., allium compounds, flavonoids, isoflavones, lignans)	

[a]The "convincing" and "probable" categories in this report correspond to the "sufficient' category of the IARC report on weight control and physical activity (4) in terms of the public health and policy implications

[b]For colorectal cancer, a protective effect of fruit and vegetable intake has been suggested by many case-control studies but this has not been supported by results of several large prospective studies, suggesting that if a benefit does exist it is likely to be modest

Importantly, there is convincing evidence that physical activity decreases the risk of colon cancer. Dietary factors that probably increase the cancer risk include high intake of preserved meat (colorectum), of salt-preserved foods and salt (stomach) and of very hot drinks and spicy food (oral cavity, pharynx and esophagus). In contrast, probably protective factors are the consumption of fruit and vegetables (oral cavity, esophagus, stomach and colorectum) and physical activity (breast cancer). According to the *World Cancer Research Fund* (*WCRF*) and the *American Institue for Cancer Research* (*AICR*), intake of foods containing fibre is also a probably protective factor against colorectal cancer, while intake of red meat and processed meat convincingly increases the risk of this type of cancer. Thus, in order to prevent cancer people should eat a healthy diet consisting of plenty of whole grains, pulses, vegetables and fruits, and limit the intake intake of high-calorie foods (foods high in sugar or fat), red meat and foods high in salt. In addition, intake of processed meat and sugary drinks should be avoided.

One of the largest difficulties to study the role of diet on cancer risk are caused by the inaccuracy of estimating dietary intake, mostly due to recall bias. Therefore, especially results from mostly available case-control studies on diet that have retrospectively assessed the food intake among people who have already developed cancer are less reliable than results from prospective cohort studies. In the latter, diet is assessed among healthy subjects, who are then followed for a long period (probably 10 years or more) before the diagnosis of cancer. Randomized controlled trials (RCTs) eliminate the bias and positive results from such studies can provide evidence that the intervention has caused changes in cancer risk. Such trials have to be very large, are expensive and can only test a specific food or dietary pattern over a few years. Thus, only few results are available from trials of dietary factors and cancer risk, while negative results do not rule out the possibility that there could be an effect at a different dose, a longer duration or if the intervention included subjects with a different age.

So far, the *European Prospective Investigation into Cancer and Nutrition* (*EPIC*) is the largest diet and disease study ever undertaken. Since 1992 it has recruited more than half a million people from 10 European countries (23 centers), which have been physically examined and completed lifestyle surveys including detailed diet questionnaires. To avoid self-evaluation bias, *EPIC* combined dietary intake data with objective dietary biomarkers assessed from blood of the participants. Results from this study will give us a much better understanding of the role of diet on cancer risk in the future.

In cancer cells, oncogenes and tumor suppressor genes regulate the metabolism of nutrients, and thus mutations in genes encoding for metabolic enzymes can contribute to the development of cancer (Fig. 1.2). There is a large variation in cancer risk between world regions and types of cancer. For example, there is a marked overall difference in incidence and mortality of colon cancer in different parts of the EU. Greece shows the lowest incidence rate and mortality from colon cancer in Europe, while Germany has the highest incidence rate and mortality being more than double compared to Greece (Fig. 1.3). Cancer risk is mainly due to environmental factors affecting the epigenome (Chap. 5) rather then due to genetic factors, as migrant studies showed that moving from a region with low risk to one with a high risk leads to the same cancer pattern of the host country within one generation. Studies with identical twins also support the role of environmental factors affecting the epigenome, since the genetic risk to develop cancer at the same site is less than 10 %.

1.4 Nutrition and Diabetes

In 2012, diabetes (type 1 and type 2) caused 1.5 million deaths worldwide (representing 4 % of all deaths from non-communicable diseases). In 2013 the worldwide number of persons with T2D was around 387 million (Sect. 10.1). However, this

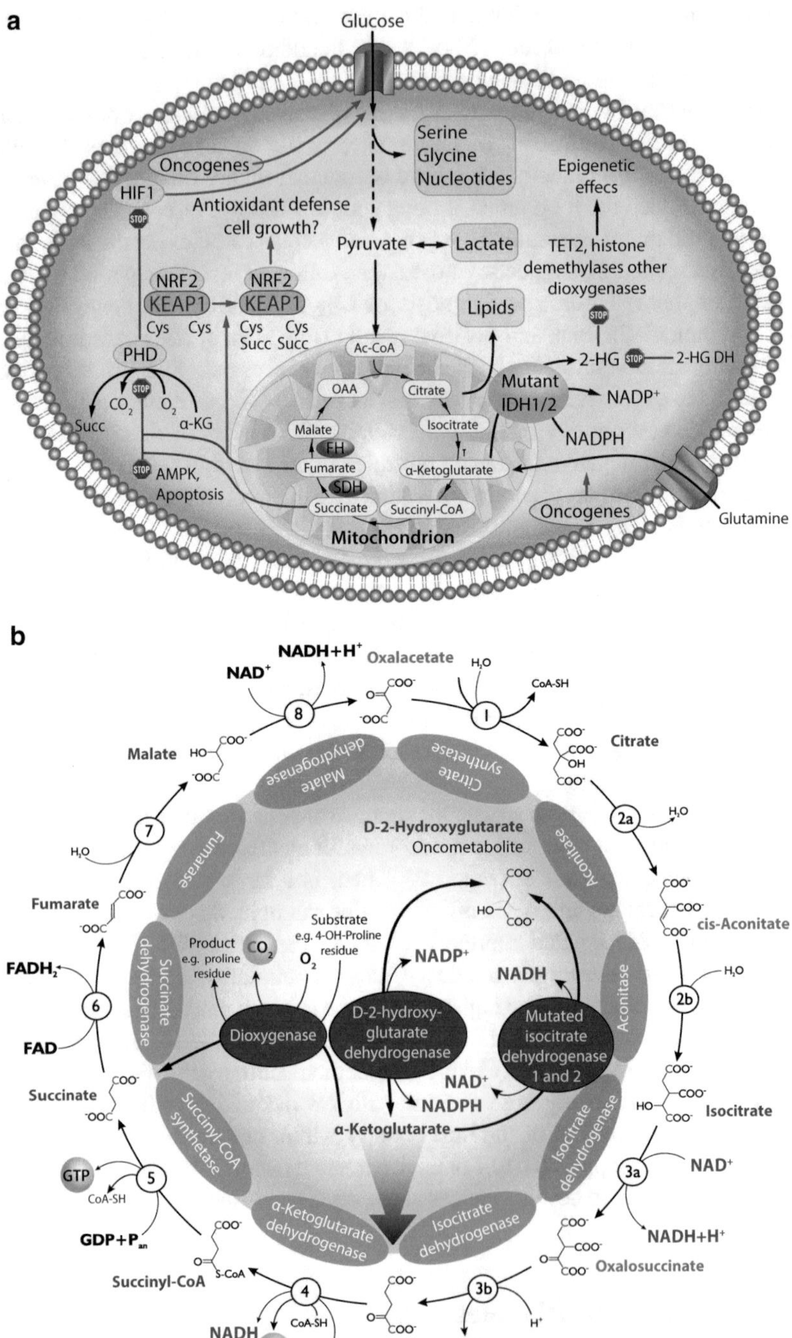

Fig. 1.2 Cancer cell metabolism. (**a**). The two major nutrients consumed by cancer cells are the amino acid glutamine and glucose, which provide precursors for (i) nucleotides, (ii) the amino acids serine and glycine and (iii) lipids. Inside mitochondria glutamine is converted into glutamate.

number is most likely underestimated, since there are many undiagnosed cases, especially in developing countries, which catch up with diseases related to overnutrition. The early stages of T2D are characterized by an overproduction of insulin as a result of insulin resistance of skeletal muscle and/or the causing increased glucose levels when the disease progresses (Sect. 9.2). The prevalence and mortality rates of T2D differ in the world. Highest prevalence is in Qatar and Saudi Arabia and lowest prevalence is in the Ukraine and Nigeria, while the total number of deaths related to T2D is highest in India and China. Countries with the highest prevalence and rates of mortality spend less money per patient (India and China) than USA, Australia, Germany and United Kingdom (Fig. 1.4). A main concern with T2D is that not only elderly but also children and adolescents develop the disease. Complications of T2D also include increased risk of blindness, kidney failure, coronary heart disease (CHD) and stroke (Sect. 10.1).

In order to prevent T2D and its complications, people should maintain a healthy body weight by being physically active for least 30 min of regular, moderate-intensity activity on most days. In addition, people should eat a healthy diet consisting of between 3 and 5 servings of fruit and vegetables per day and reduce sugar and SFA intake. These recommendations are based on the fact that there is convincing evidence that weight gain, visceral fat and physical inactivity increases the risk of T2D (Table 1.3). The dietary factor that probably decreases the risk of T2D is non-starch polysaccharide, which is the main constituent of dietary fiber. In contrast, the intake of SFAs probably increases the risk of T2D. The molecular mechanisms behind the development of T2D and the possible role of inflammation and some dietary factors are further described in Chaps. 9 and 10.

1.5 Nutrition and Cardiovascular Diseases

CVDs are the major contributor to the global burden of disease among the non-communicable diseases and in 2012 caused 17.5 million deaths, i.e. 46 % of non-communicable disease deaths worldwide (Sect. 11.2). Tobacco use, physical inactivity and unhealthy diet are responsible for about 80 % of CHD and cerebrovascular disease, i.e. heart attack and stroke. Ischemic or CHD, i.e. the failure to supply blood to the heart, is the major cause of CVD deaths (42.5 %), while cerebrovascular disease, i.e. the failure to supply blood to the brain, causes 35.5 % of the

Fig. 1.2 (continued) (**b**). Glutamate is the major source of the TCA cycle intermediate α-ketoglutarate that is a substrate of dioxygenases, such as prolyl hydroxylases, histone demethylases (HMTs) and 5-metylcytosine hydroxylases modifying proteins and DNA. Thus, α-ketoglutarate is an essential component of cell signaling and epigenetic networks. Oncogenic signaling regulates uptake and catabolism of glucose and glutamine. Presumed metabolic tumor suppressors, such as kelch-like ECH-associated protein 1 (KEAP1), fumarate hydratase (FH), succinate dehydrogenase (SDH) and D-2-hydroxyglutarate dehydrogenase (D2HGDH) as well as oncogenes, such as isocitrate dehydrogenase (IDH) 1 and 2, control the abundance of key metabolites that regulate signaling functions. These signaling activities contribute to malignant transformation in cancer cells

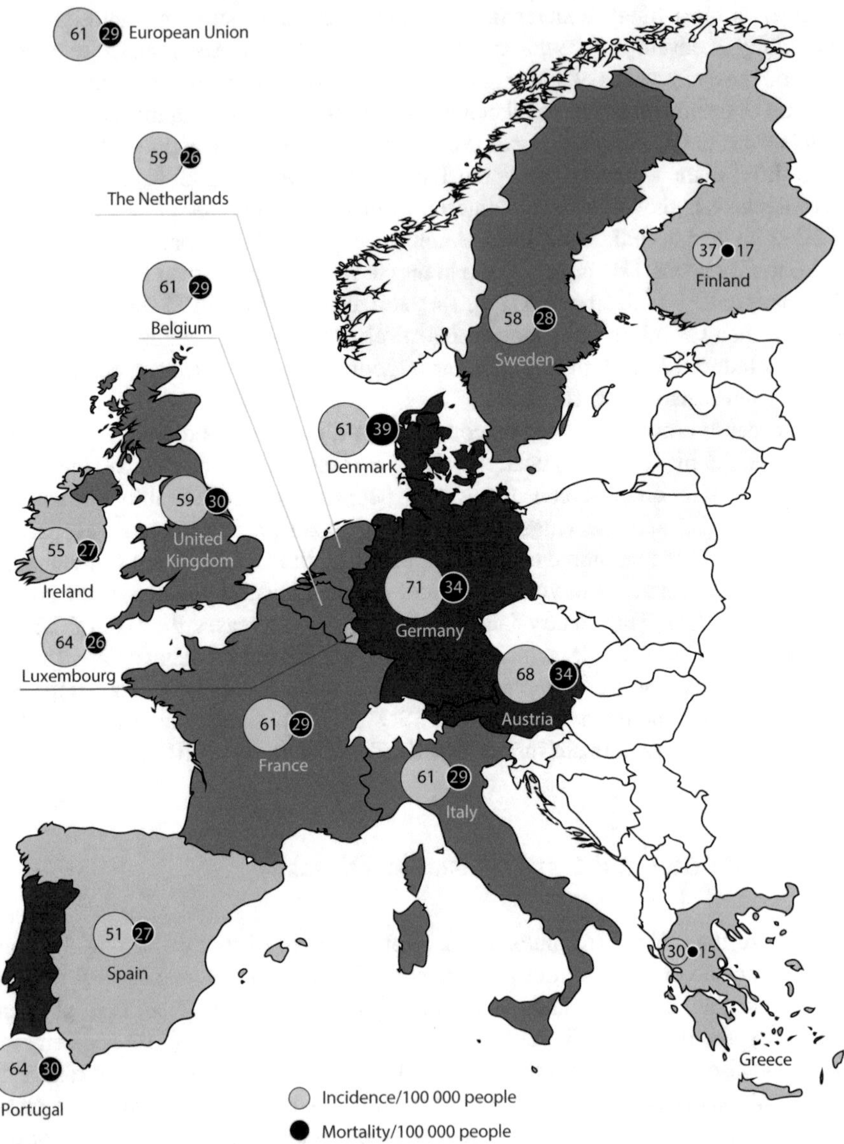

Fig. 1.3 Rates of colorectal cancer. Incidence and disease-specific mortality per 100,000 people in 15 EU countries. Data were collected from the *EUCAN* website (http://eu-cancer.iarc.fr/EUCAN/Default.aspx), which is administered by the *International Agency for Research on Cancer*. EUCAN disseminates country-specific information on cancer burden, derived from data collected by the national cancer registries and from the *WHO* mortality database

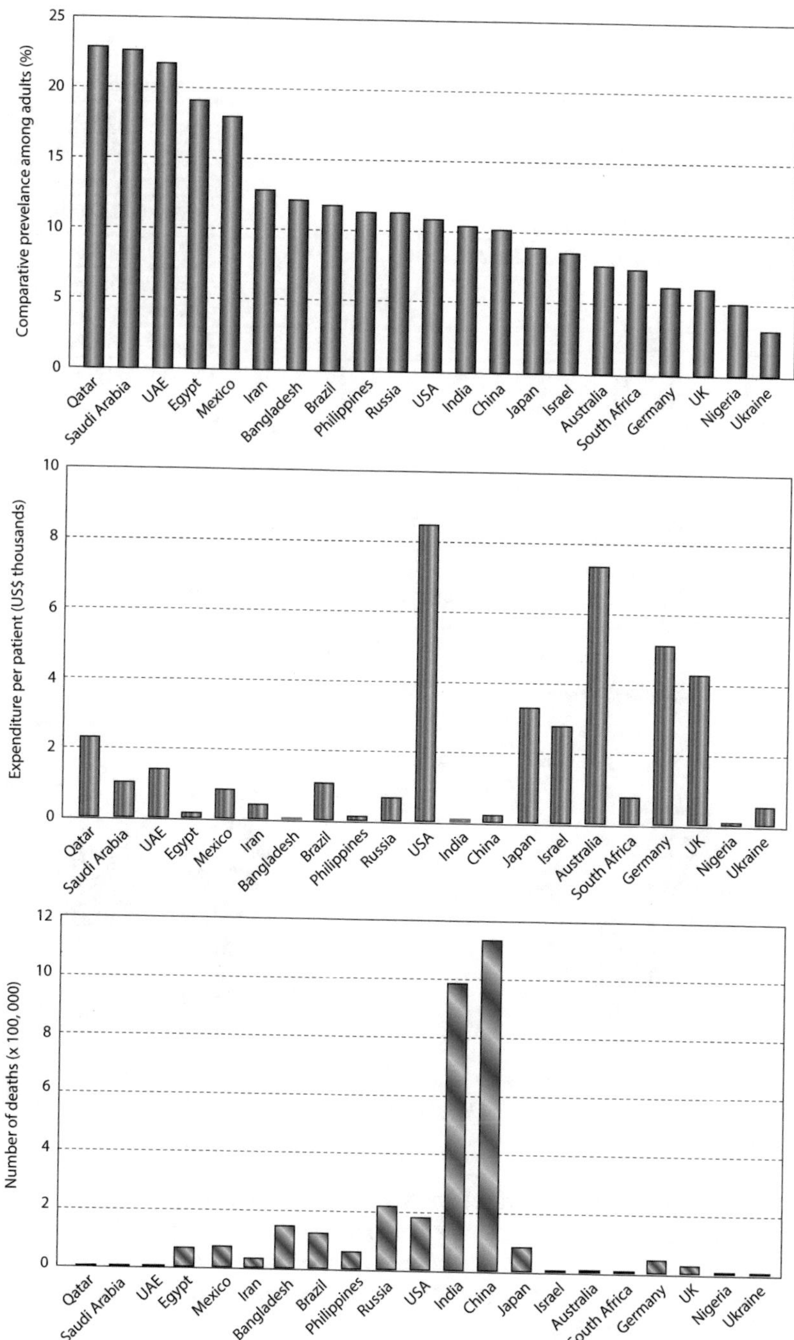

Fig. 1.4 Prevalence, total expenditure per patients and number of diabetes-related deaths selected countries in 2011

Table 1.3 Overview of lifestyle factors and risk of developing T2D

Evidence	Decreased risk	No relationship	Increased risk
Convincing	Voluntary weight loss in overweight and obese people		Overweight and obesity
			Abdominal obesity
			Physical inactivity
	Physical activity		Maternal diabetes[a]
Probable	Non-starch polysaccharides		Saturated fats
			Intrauterine growth retardation
Possible	ω-3 fatty acids		Total fat intake
	Low glycemic index foods		Trans fatty acids
	Exclusive breastfeeding[b]		
Insufficient	Vitamin E		Excess alcohol
	Chromium		
	Magnesium		
	Moderate alcohol		

[a]Includes gestational diabetes
[b]As a global public health recommendation, infants should be exclusively breastfed for the first six months of life to achieve optimal growth, development and health

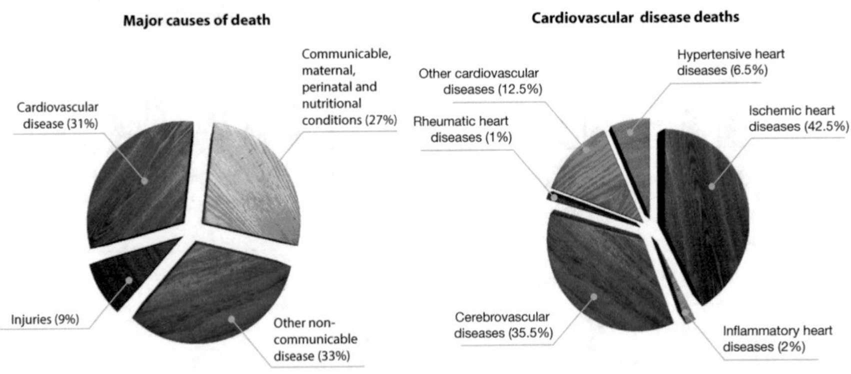

Fig. 1.5 CVD accounts for nearly one-third of deaths worldwide. Ischemic heart diseases are caused primarily by clogged arteries (atherosclerosis) and are responsible for most of the CVD deaths

CVD deaths (Fig. 1.5). Atherosclerosis is the basic pathophysiological lesion of CVD, which tends to occlude the arteries to a varying extent (Sect. 11.2). A variety of cell types and lipids are involved in the pathogenesis of atherosclerotic plaques and arterial thrombosis. Moreover, also in the progression of atherosclerosis nutrition plays an important role.

There is convincing evidence that consumption of (i) fruits, berries and vegetables, (ii) fish and fish oil (containing the marine ω-3 fatty acids eicosapentaenoic acid (EPA) and docosahexaenoic acid (DHA)), (iii) foods rich in the essential fatty

Table 1.4 Overview of lifestyle factors and risk of developing cardiovascular disease

Evidence	Decreased risk	No relationship	Increased risk
Convincing	Regular physical activity, linoleic acid, fish and fish oils, vegetables, fruit and berries, potassium, low to moderate alcohol intake	Vitamin E supplements	Myristic and palmitic acids
			Trans fatty acids
			High sodium intake
			Overweight
			High alcohol intake (for stroke)
Probable	α-Linolenic acid	Stearic acid	Dietary cholesterol
	Oleic acid		Unfiltered boiled coffee
	Non-starch polysaccharides		
	Wholegrain cereals, nuts (unsalted) plant sterols/stanols, folate		
Possible	Flavonoids		Fats rich in lauric acid
	Soy products		Impaired fetal nutrition
			Beta-carotene supplements
Insufficient	Calcium		Carbohydrates
	Magnesium		Iron
	Vitamin C		

acid linoleic acid (LA) and potassium as well as (iv) physical activity and low to moderate alcohol intake all contribute to the reduction of CVD risk (Table 1.4). In addition, the essential ω-3 fatty acid α-linolenic acid (ALA), oleic acid, fibre, whole grain cereals, nuts, plant sterols and folate reduces the risk of CVD. There is also convincing evidence that intake of SFAs (myristic acid and palmitic acids), *trans*-fatty acids, sodium, overweight and high alcohol intake increases the risk of CVD, while dietary cholesterol and unfiltered boiled coffee increases the risk of CVD.

Dietary reference values are mainly based on requirements from population groups and not from individuals. However, enormous variability exists in individual responses to diet and food components that affect overall health. Both genetic and environmental factors influence the individual's response. Discoveries underpinning this variability will lead to advances in personalized nutrition as well as in improved health and food policies, including dietary reference intakes for nutrient needs and future dietary recommendations. A top priority for future nutrition research is a better understanding of the variability in metabolic responses to diet and food. Cellular and molecular characterization of disease phenotypes is therefore crucial to understand the role of food components in disease prevention and treatment, which is one of the key concepts of nutrigenomics.

The pathogenesis of T2D, certain cancers, CVD and neurodegenerative diseases are closely linked with inflammation (Fig. 1.6). Moreover, obesity and the role of dietary factors on immune function is crucial to understand the underlying mecha-

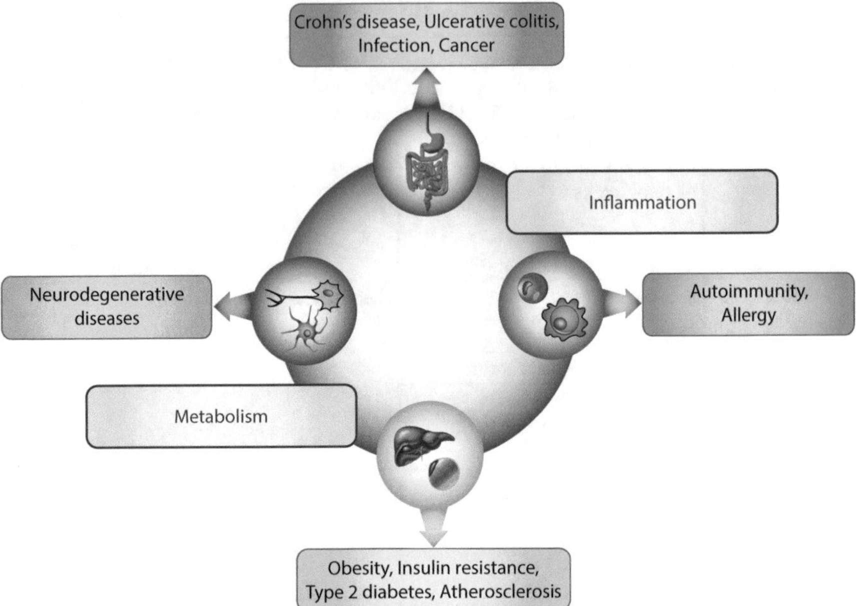

Fig. 1.6 Metabolic and immune-mediated pathologies as key driver processes of diseases in various target organs

nisms how diet can prevent diseases. The role of genetic susceptibility will also be important to determine who will benefit from specific interventions.

Cancer and obesity are examples of non-communicable diseases, in which inflammation is the underlying cause of the disease (Sects. 7.4 and 8.1). Several studies have shown an elevated risk of different types of cancer, such as pancreatic, prostate, breast and colon cancer, is concomitant with high BMI. White adipose tissue (WAT) is an important endocrine and metabolic organ consisting of both lipid-loaded adipocytes and a stromal-vascular fraction, which contains pre-adipocytes, macrophages, other inflammatory cells and endothelial cells (Fig. 1.7a). The latter fraction may represent the cell population linking obesity to cancer. Obesity increases the size of adipocytes (hypertrophy) and number of adipocytes (hyperplasia) and is accompanied by infiltration of macrophages in the adipose tissue (Sect. 8.5). Inflammatory responses that are triggered by obesity are characterized by elevated levels of circulating pro-inflammatory cytokines and acute phase proteins, such as C-reactive protein (CRP). In addition, increased release of pro-inflammatory adipokines, such as leptin, interleukin (IL) 6, serpin peptidase inhibitor, clade E (SERPINE1, also called plasminogen activator inhibitor 1) and tumor necrosis factor (TNF), and reduced release of anti-inflammatory adipokines, such as adiponectin, is associated with obesity (Fig. 1.7a).

The link between obesity and cancer initiation as well as the molecular mechanisms underlying how obesity converses normal epithelial cells to tumor cells is not completely understood (Fig. 1.7b). In addition to low-grade inflammation and the

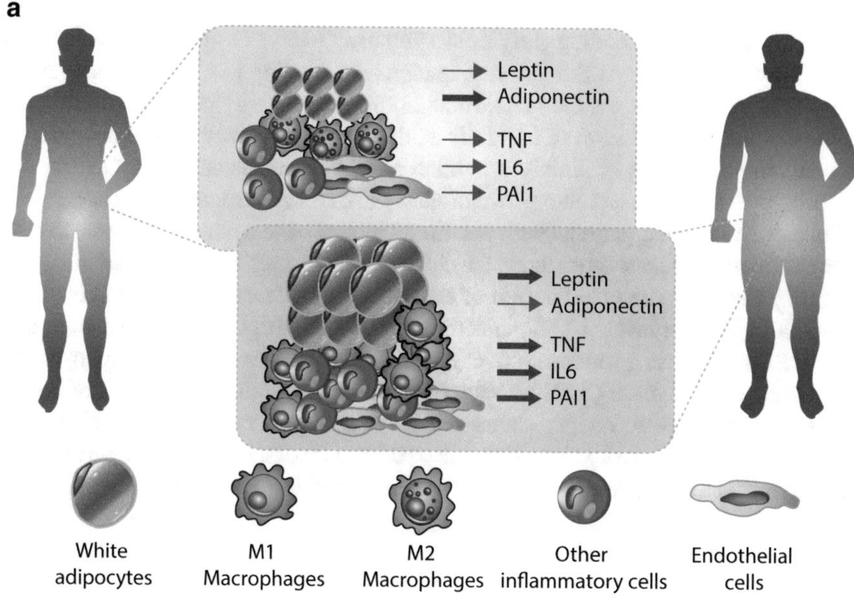

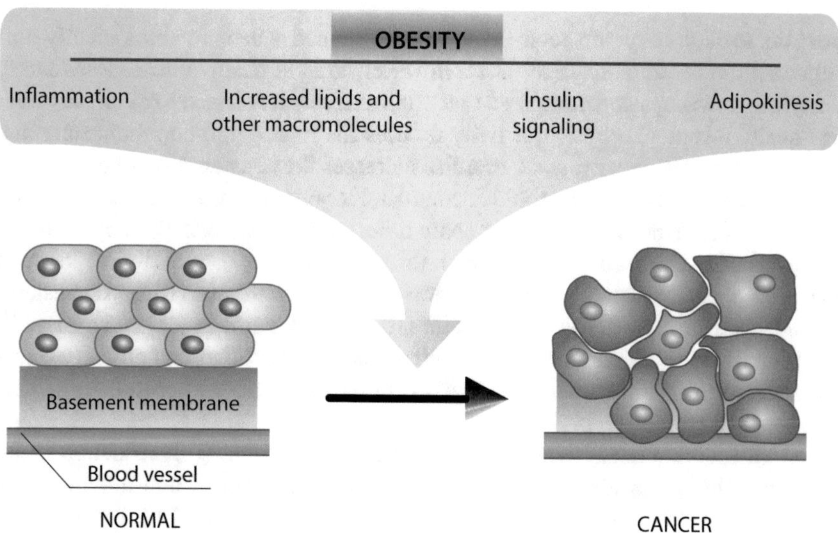

Fig. 1.7 Obesity, inflammation and risk of disease. (**a**). WAT is an endocrine and metabolic organ consisting of lipid-loaded adipocytes and pre-adipocytes, macrophages, other inflammatory cells and endothelial cells. In subjects with normal weight, the adipose tissue secretes high levels of adiponectin. During weight gain, WAT expands, which mediates the infiltration of macrophages and other inflammatory cells and leads to the secretion of the cytokine TNF from macrophages. Furthermore, the secretion of IL6, SERPINE1 (also called PAI1) and leptin is also increased. (**b**). Elevated inflammation, increased availability of lipids and other macromolecules, impaired insulin signaling and changes in adipokine signaling all contribute to the conversion of epithelial cells to an invasive tumor. Although all of these pathways can contribute to cancer in certain circumstances, it remains unclear whether they are predominantly required for the development of cancer in obese humans

release of inflammatory cytokines, lipid metabolism is altered in cancer cells. For example, the gene encoding fatty acid synthase (*FASN*) is up-regulated by fatty acids, and the activity of this enzyme is increased leading to elevated endogenous fatty acid synthesis in cancer cells. In addition, high activity of lipolytic enzymes increases the concentration of free fatty acids (FFAs) either acting as tumor growth stimulating molecules or simply providing energy to the cancer cells. Obesity can also influence insulin signaling, which may provide further energy to cancer cells (Chap. 9). Moreover, patients with insulin resistance have a poorer response to cancer treatment or bear a more aggressive cancer phenotype. Elevated insulin signaling may also promote proliferation of cancer cells. Furthermore, elevated levels of leptin and reduced levels of adiponectin stimulate tumor growth. Adiponectin acts via transmembrane proteins activating the kinase adenosine monophosphate-activated protein kinase (AMPK, Sect. 6.3). AMPK is a critical regulator of proliferation in response to energy status and plays a role in the regulation of growth arrest and apoptosis. Adiponectin can also activate the nuclear receptor peroxisome proliferator-activated receptor (PPAR) α that controls fatty acid oxidation (Sect. 3.3).

1.6 Impact of Exercise

From the evolutionary perspective, an exercise-trained state is the biologically normal condition for humans (Sect. 2.1). However, in industrial countries a sedentary lifestyle is nowadays so widespread that often exercise is referred to as already having "health benefits". Physical activity reduces the risk of non-communicable diseases and prevents obesity, since exercise increases the consumption of energy, i.e. it burns off the body fat that would accumulate. Regular exercise, i.e. the voluntary activation of skeletal muscle during spare time, sports and work, promotes cardiovascular health. Exercise is beneficial for the profile of serum lipids, since it decreases plasma triacylglycerols and low-density lipoprotein (LDL) cholesterol and increases of high-density lipoprotein (HDL) cholesterol (Sect. 11.3). In addition, physical activity also has an anti-inflammatory effect that can protect against low-grade chronic inflammation-associated diseases, such as atherosclerosis, T2D and cancer (Fig. 1.8a).

The anti-inflammatory effect of exercise is mediated via (i) a reduction in fat mass, (ii) a higher production and release of myokines, i.e. of anti-inflammatory cytokines of skeletal muscle, and (iii) the reduced expression of Toll-like receptors (TLRs). TLRs are pattern recognition receptors (PRRs) on the surface of monocytes and macrophages that detect pathogens and initiate the innate immune response (Sect. 7.2). These anti-inflammatory effects inhibit the production of pro-inflammatory cytokines (Fig. 1.8b). Exercise also reduces the number of pro-inflammatory monocytes and increases that of circulating regulatory T cells (T_{REG}). T_{REG} are a specialized sub-population of T cells suppressing the activation of the immune system (Sect. 7.1). Regular exercise results in a reduced fat mass and can

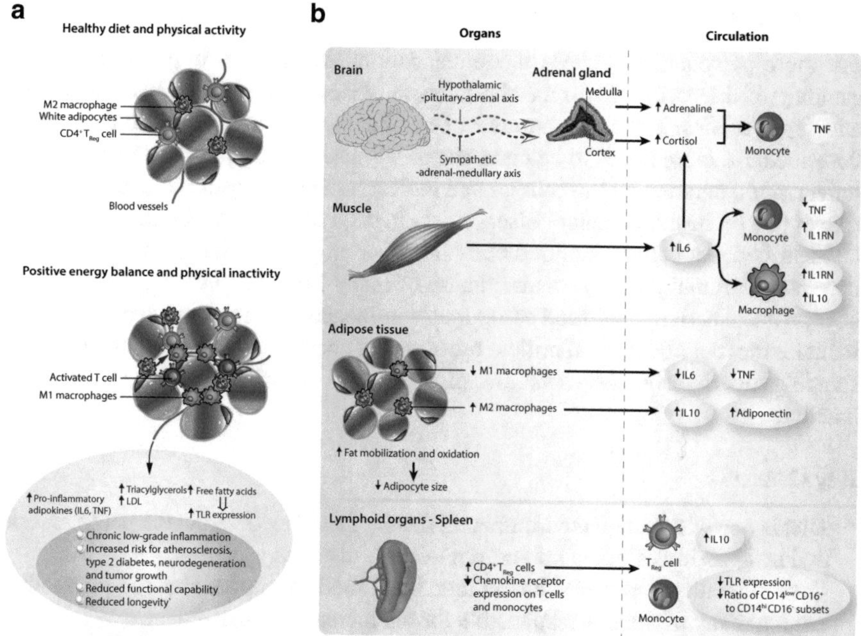

Fig. 1.8 Risk of disease by inflammation, exercise and diet. (**a**). The anti-inflammatory phenotype of adipose tissue is maintained by healthy diet and physical activity. This is marked by small adipocyte size and the presence of anti-inflammatory macrophages (M2-type). In contrast, a positive energy balance and physical inactivity lead to the accumulation of fat and the infiltration of adipose tissue with pro-inflammatory macrophages (M1-type) and T cells. These M1 macrophages release pro-inflammatory adipokines, such as TNF and IL6, which cause low-grade inflammation, thus promoting the development of chronic diseases. The lack of exercise also impairs the blood lipid profile stimulating the development of atherosclerosis. (**b**). Exercise leads to reduced mass of adipose tissue mass, smaller adipocyte size, less macrophage infiltration and a switch from M1 to M2 macrophages. Furthermore, exercise increases the release of the hormones cortisol and adrenaline from the adrenal cortex and medulla, respectively. Both these hormones inhibit the release of TNF by monocytes. IL6 produced by contracting of skeletal muscles stimulates the release of IL1RN from monocytes and macrophages, i.e. the levels of this anti-inflammatory cytokine in the circulation increases. Exercise mobilizes T_{REG} cells – which are a major source of the anti-inflammatory cytokine IL10 – from the spleen into circulation

reduce the infiltration of macrophages to adipose tissue and a switch from M1 to M2 macrophages (Sect 7.4). This leads to an increase in adiponectin levels and a decrease in pro-inflammatory adipokines (IL6, TNF and leptin) in the circulation. During and following exercise the active skeletal muscle markedly increases both cellular and circulating levels of IL6. This transient increase of IL6 during exercise leads to an increase in the anti-inflammatory cytokines IL10 and IL1 receptor antagonist (IL1RN). Exercise also influences the CNS, since impulses from the brain and contracting muscles elevate plasma cortisol and adrenaline production in adrenal glands. These hormones suppress inflammation by decreasing the pro-inflammatory cytokine production by monocytes and macrophages.

Future View

For more personalized dietary recommendations we need to improve the understanding of the variability in metabolic responses caused by diet, physical activity and (epi)genetics. Even if a number of mechanisms of nutrient actions are well described, far more needs to be understood. Disease phenotypes have to be characterized more detailed on the cellular and molecular level, in order to understand the role of diet on health benefit or disease risk. In this context, new biomarkers have to be identified that better monitor dietary intake or physical activity. Since low-grade chronic inflammation is the central cause of many lifestyle diseases, more insight how physical activity and food components influence inflammation is needed. This includes the identification of critical processes of energy homeostasis and an understanding how these pathways are disrupted in disorders that are related to inactivity.

Key Concepts

- Diet is one of the main environmental factors that is involved in the pathogenesis and progression of many non-communicable diseases.
- Overall quality of diet and not an individual food component but the interaction among many of them is responsible for the increased disease risk.
- Carbohydrates (digested to glucose), proteins (digested to amino acids) and fats (digested as fatty acids and cholesterol) are the main macronutrients that support the body with energy or are stored when there is excess of energy. Disturbances in the metabolism of these macronutrients can cause diseases.
- In a healthy diet, the acceptable macronutrient distribution ranges of carbohydrates, protein and fat are 45–65 E%, 10–35 E% and 20–35 E%, respectively.
- Obesity and high consumption of alcohol and of some forms of salted and fermented fish increases the risk of certain cancers, while physical activity decreases the risk of colon cancer.
- The complex interaction between genetic variation, individual metabolic characteristics and diet significantly contributes to cancer.
- Different dietary factors can both increase or reduce cancer risk, but inaccuracies in estimating dietary intake makes it difficult to define relationships between specific nutrient-related factors and cancer risk.
- Weight gain, visceral fat and physical inactivity increase the risk of T2D.
- Dietary fibre and regular physical activity reduce the risk of obesity, while high intake of energy-dense foods increases the risk.
- Consumption of fruits, berries and vegetables, fish and fish oil, foods rich in the essential fatty acid LA and potassium, as well as physical activity and low to moderate alcohol intake, reduce the risk of CVD, while intake of SFAs, *trans*-fatty acids and sodium each increase the risk.
- Inflammation is etiologically linked to the pathogenesis of cancer, obesity, T2D and CVD.
- Elevated levels of circulating inflammatory cytokines and acute phase proteins characterize obesity-triggered inflammation.

- Regular exercise promotes cardiovascular health by decreasing the concentration of plasma triacylglycerols and LDL-cholesterol and by increasing the HDL-cholesterol levels.
- Physical activity has an anti-inflammatory effect that protects against low-grade chronic inflammation-associated diseases.

Additional Reading

Calder PC, Ahluwalia N, Brouns F, Buetler T, Clement K, Cunningham K, Esposito K, Jönsson LS, Kolb H, Lansink M, Marcos A, Margioris A, Matusheski N, Nordmann H, O'Brien J, Pugliese G, Rizkalla S, Schalkwijk C, Tuomilehto J, Wärnberg J, Watzl B, Winklhofer-Roob BM (2011) Dietary factors and low-grade inflammation in relation to overweight and obesity. Br J Nutr 106(Suppl 3):S1–S78

de Roos B (2013) Personalised nutrition: ready for practice? Proc Nutr Soc 72:48–52

Ezzati M, Riboli E (2013) Behavioral and dietary risk factors for noncommunicable diseases. N Engl J Med 369:954–964

Gleeson M, Bishop NC, Stensel DJ, Lindley MR, Mastana SS, Nimmo MA (2011) The anti-inflammatory effects of exercise: mechanisms and implications for the prevention and treatment of disease. Nat Rev Immunol 11:607–615

Hawley JA, Hargreaves M, Joyner MJ, Zierath JR (2014) Integrative biology of exercise. Cell 159:738–749

Hensley CT, Wasti AT, DeBerardinis RJ (2013) Glutamine and cancer: cell biology, physiology, and clinical opportunities. J Clin Invest 123:3678–3684

Lumeng CN, Saltiel AR (2011) Inflammatory links between obesity and metabolic disease. J Clin Invest 121:2111–2117

Chapter 2
Human Genomic Variation

Abstract Due to partial isolation of human populations during history, their genetic variation is geographically diverted. Positive natural selection, i.e. the force that drives the increase in prevalence of advantageous traits, has played a central role in human evolution. Genetic differences between human populations are most pronounced in tissues, such as the skin, the intestinal tract or the immune system, that are directly affected by the environment. This led not only to obvious differences in skin color among the populations, but also in different resistance to diseases and diversity in dietary intake, such as the ability to digest milk sugar (lactose). The genetic basis of the variation of human populations and individuals has recently been studied and catalogued by large consortia, such as the *HapMap Project* and the *1000 Genomes Project*. They obtained data via genome-wide genotyping and whole genome sequencing of 2504 subjects and thus allow the study and analysis of the relation of human genomic variation and disease risk.

In this chapter, we will briefly describe the genetic adaption of the anatomically modern human to new geographic and climatic environments in Asia and Europe and the challenges provided by the shift from hunters and gatherers to farmers (Chap. 4). We will discuss how complex phenotypic traits influence the risk to develop diseases, such a T2D and CVD (Chaps. 10 and 11). Each complex disease is based on dozens to hundreds of gene variants, such as single nucleotide polymorphisms (SNPs) and structural variants, such as copy number variations (CNVs). We will describe how the *HapMap Project* and the *1000 Genomes Project* map these genetic variants in different human populations. In this context, we will discuss how whole genome sequencing can result in the identification of rare SNPs that significantly contribute to complex traits and diseases.

Keywords Human evolution • Human populations • Single nucleotide variants • Copy number variants • Haplotype blocks • Next-generation sequencing • *HapMap Project* • Genome-wide association studies • *1000 Genomes Project*

C. Carlberg et al., *Nutrigenomics*, DOI 10.1007/978-3-319-30415-1_2

2.1 Migration and Evolutionary Challenges of the Modern Human

Approximately 200,000 years ago the anatomically modern human (*Homo sapiens sapiens*) developed in East Africa. The main characteristic of this modern human is a superior locomotive ability that is essential for encountering of predators and food procurement. Some 60–80,000 years ago, some of these modern humans started to migrate to Asia and Europe and there replaced, without significant interbreeding (less than 4 % common genome sequence) already prevalent archaic *Homo sapiens* species, such as the Neanderthals (Fig. 2.1). Due to their new environments our ancestors were exposed to a number of divergent selective pressures, such as thermoregulation in colder climates, tolerance to hypoxia at high altitude and light skin pigmentation in regions with lower levels of ultraviolet (UV)-B radiation (Sect. 4.2). Moreover, within this period humans were exposed to periodic cycles of feasts and famines, which let certain gene variants, the so-called "thrifty genes" (Sect. 10.4), evolve that regulate efficient storage and utilization of energy from food.

Some 10,000 years ago our ancestors started to give up their hunter and gatherer habit and became farmers. This significant lifestyle change was associated with distinct foods, such as cereals and milk (Sect. 4.4). The improved nutrition supply allowed higher population densities but was compromised with an increased load with infectious diseases, many of which were acquired from domesticated animals. Both diet change and immunological challenge caused dominant evolutionary pressure and rather rapid genetic adaption. The phenotypic consequences of these adaptive genetic variants did not only shape the biological variation but also the health and disease risk of the more than seven billion humans presently living on Earth. However, the by far most drastic lifestyle change happened during the last 100–200 years, i.e. in less than 10 generations (Sect. 1.1). Industrial revolution invented machines, such as trains, automobiles and aeroplanes, the use of which caused a

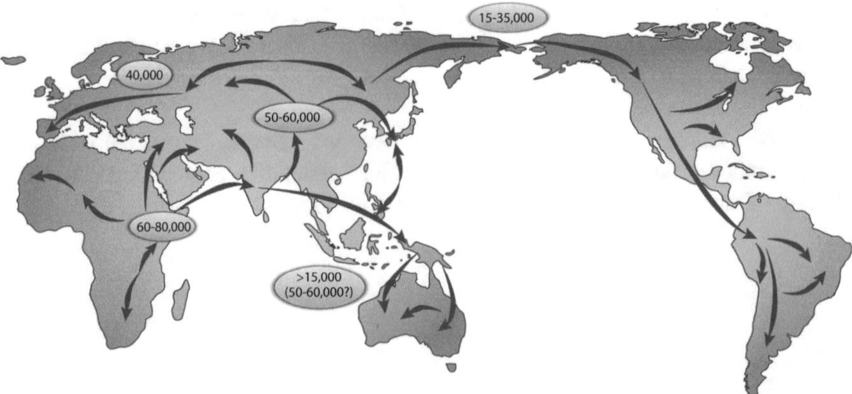

Fig. 2.1 The migration of modern *Homo sapiens*. Approximately 60–80,000 years ago the spread from East Africa over the rest of the continent was followed by an expansion from the same area to Asia, probably by both a southern and northern route some 40–60,000 years ago. Oceania, Europe and America were settled from Asia in that order

significant reduction in the physical activity of most humans. In parallel, the world-wide spreading of easy available, highly processed food led to overnutrition of the majority of the human population. In Chap. 1 we already started to discuss the significant increase in obesity and the resulting increase in the rates of cancer, T2D and CVD. Worldwide it took several thousand years, i.e. clearly more than 100 generations, to turn most of the human population from hunters to farmers, but less than 50 years to be preferential users of cars, supermarkets and fast-food. This means that the human population had simply no time to adapt genetically to the rapidly changing "obesogenic environment". In the context of an inactive lifestyle combined with energy-dense foods, the gene alleles that had been initially evolved for an efficient energy storage and physical mobility, increased the risk for chronic diseases, such as T2D or CVD.

2.2 Diversity of Human Populations

When humans became distributed to the different continents, new gene variations could not be spread to all humans. Therefore, during the last 50,000 years, when humans accumulated further mutations, population-specific alleles of human genes evolved (Fig. 2.2). This resulted in phenotypic differences of the different geographic populations concerning skin color, body height and facial features. However, there are no absolute genetic differences between the populations on the different continents. For example, there is no single nucleotide difference that may be used as a marker, in order to distinguish, in general, Africans from Eurasians. In contrast, the different features in the physiognomy of the populations in the different continents are based on a multitude of gene loci, which carry alleles that vary in most populations. This implies that a property (in population genetics often referred to as a "trait"), such as skin color, can change rather rapidly, when the allele frequencies shifts at the loci that contribute to this trait (Box 2.1). Furthermore, we have to remind that some 500 years ago, when shipping and navigation over the oceans became possible, a large migration started, which caused significant population mixtures, particularly in the Americas, but also in other continents.

Box 2.1 Selection in Evolution

Phenotypic traits are based on gene alleles. During evolution these traits and alleles are under natural selection (in contrast to "selective breeding" in domesticated animals or plants) and segregate within a population. Individuals with advantages (referred to as "adaptive traits") tend to be more successful in reproduction, i.e. they contribute with more offspring to the following generation than others. Due to inheritance from one generation to the other, the selection process will increase the prevalence of the adaptive traits. This is called "positive selection". Under persistent selection pressure these adaptive traits, step by step, may become universal to the population, i.e. they will have

(continued)

Box 2.1 (continued)
"evolved". Factors fostering selection (referred to as "evolutionary pressures") include, for example, limits on resources, such as nutrition, or threats, such as diseases. However, adaptive traits not always will become prevalent within a population. Gene frequency alteration in a population can also happen via a "genetic drift" of genes that are not under selection. In this context, even a deleterious allele may become universal to the members of a population, for example, under the influence of a weak selection or in small populations.

Hundreds of complex phenotypic traits determine how humans look and behave as well as their risks to develop certain diseases. Each complex phenotype is based on dozens to hundreds of gene variants and environmental influences. Whole genome sequencing has indicated that every human individual carries, on average, approximately four million genetic variants (Sect. 2.4) covering about 12 Mb of sequence (0.3 % of all). Most of these genetic variants are neutral, i.e. they do not contribute to phenotypic variation or disease risk, and achieved significant frequencies in the human population simply by chance. Nevertheless, there are a number of common variants with a small to modest effect size that have a dominant role in common complex traits. In addition, the sum of rare, high-penetrance variants is of significant influence. For example, the trait "body height" is dependent on at least 180 gene loci, i.e. it is a prime example of a complex trait (for further examples see Chaps. 8, 10 and 12). In Europe, this trait has changed significantly within only a few generations (in average by 10 additional centimeters) under the environmental trigger of improved quality and quantity of nutrition.

During evolution of the anatomically modern human up to 10 % of all protein-coding genes, i.e. some 2000 genes, may have been affected by positive selection. In particular, the immune system, the digestive tract and the skin (including hair, sweat glands and sensory organs) had been susceptible to positive selection (Sect. 4.2). This is due to the fact that these organ systems are far more in contact with the environment than other parts of the body. For example, variants of the innate and adaptive immune system (Box 2.2), such as genes encoding for membrane immune receptors, had been under special positive selection by pathogenic microbes. The most recent example of this is the C-C chemokine receptor 5 (CCR5), which is essential for the entry of the human immunodeficiency virus (HIV) 1 into T cells. A 32 bp deletion in the *CCR5* gene protects its carriers from HIV infection, and this mutation is currently becoming positively selected in populations where HIV1 infections occur on larger scale. Moreover, some alleles that were introduced into modern humans through interbreeding with archaic (meaning nowadays extinct) human species seem to have been positively selected. For example, a Neanderthal

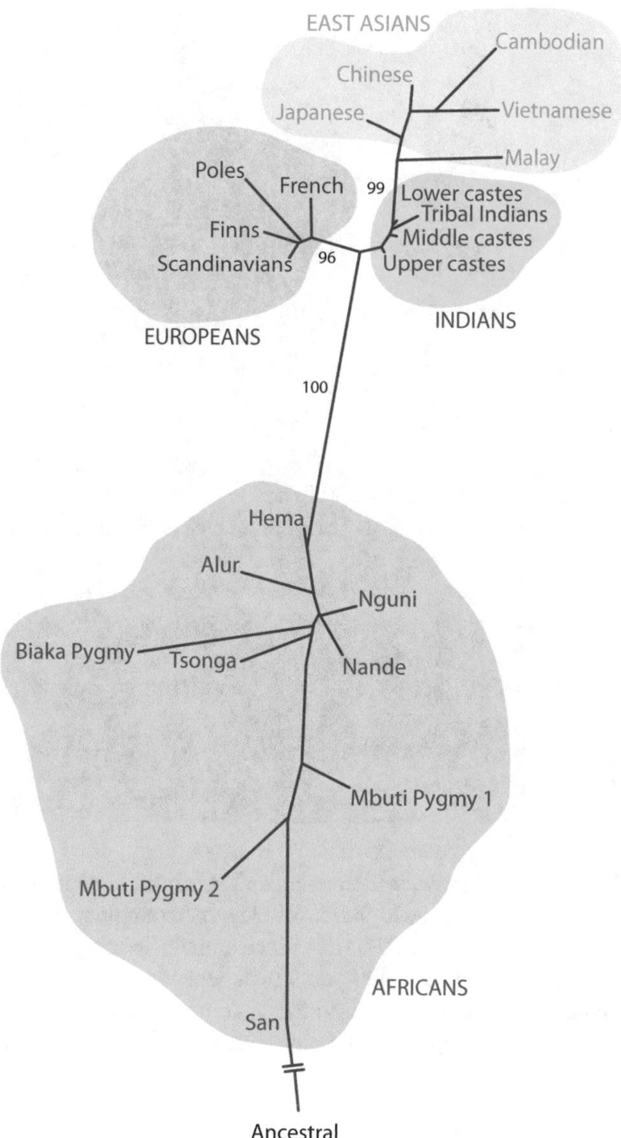

Fig. 2.2 A network of population similarities. Based on the frequencies of insertions of *Alu* short interspersed nuclear elements (SINEs) the network is rooted using a hypothetical ancestral group that lacks the *Alu* insertions at each locus. Bootstrap values are shown (as percentages) for main internal branches. Populations tend to cluster according to their geographic distance from each other. This is to be expected, as geographically distant populations were less likely to exchange migrants throughout human evolutionary history. Moreover, the African populations are more diverse, which is consistent with the fact that anatomically modern humans lived in Africa already since approximately 200,000 years. The observation that the largest genetic distances are found between African and non-African populations is another indication that the root of the tree, i.e. the origin of *Homo sapiens sapiens*, is in Africa

Box 2.2 The Innate and Adaptive Immune System
In general, the immune system is a system of biological structures, such as lymph nodes, cell types, such as monocytes, macrophages, T and B lymphocytes (i.e. cellular immunity), and proteins, such as antibodies (i.e. humoral immunity), that protect the organism against infectious diseases. The immune system detects a wide variety of molecules, known as antigens, of potential pathogenic origin, such as on the surface of microbes, and distinguishes them from the organism's own healthy tissue. In mammals, such as in humans, the immune systems functions can be classified into the innate immune system and the adaptive immune system. Innate immunity is evolutionary older, bases on the cell type's monocytes, macrophages, neutrophils and natural killer cells, and uses destructive mechanisms against pathogens such as phagocytosis, with the support of anti-microbial peptides from the complement system. Adaptive immunity applies more sophisticated defense mechanisms, in which T and B cells use highly antigen-specific surface receptors, such as T cell receptors and B cell receptors, the latter finally turning into secreted antibodies. Moreover, after an initial specific response to a pathogen the adaptive immune systems create an immunological memory that leads to an enhanced response to subsequent encounters with that same antigen. Peripheral blood mononuclear cells (PBMCs) are immune cells found in the blood, i.e. primarily the cell types monocytes and lymphocytes, which are easily accessible. Since they are exposed to nutrients and metabolic hormones in the circulation these cells are a good model system to detect metabolic and dietary changes

variant of the *SLC16A11* gene, which encodes for a lipid transporter in the endoplasmic reticulum (ER), reached high frequencies, for example in native Americans, being associated with increased T2D risk. Since archaic human populations outside of Africa experienced some 300,000 years of independent accumulation of mutations, their maximal 4 % contribution to the gene pool of the modern human may have an overproportional impact on the physiology of present-day people.

2.3 Genetic Variants of the Human Genome

The reference haploid sequence of the human genome (Box 2.3) was released in 2001 by the first "big biology" project, the *Human Genome Project* (Box 2.4), and reflects the assembly of sequences derived from a few donors. SNPs are variants of the reference sequence where exactly one nucleotide (A, T, G or C) is altered (Fig. 2.3). In contrast, structural variants of the genome mostly affect more than one nucleotide. These can be insertion-deletion variants (indels), where in most cases only a few bases are added or removed, respectively, but there are also indels of up to 80 kb in length. Indels that are not multiples of 3 bp in length and are located

Box 2.3 The Human Genome

The human genome is the complete sequence of the anatomically modern human (*Homo sapiens sapiens*). With the exception of germ cells, i.e. female eggs and male sperm, each human cell contains a diploid genome formed by more than 2x3 billion bp that is distributed on 2x22 autosomal chromosomes and two X chromosomes for females and a XY chromosome set for males. In addition, every mitochondrium contains 16.6 kb mitochondrial DNA. The haploid human genome encodes for some 20,500 protein-coding genes and about the same number of non-coding RNA (ncRNA) genes. The coding sequence covers less than 2 % of the human genome, i.e. the vast majority of the genome is non-coding and seems to have primarily regulatory function.

The reference human genome is accessible via different browser websites (for example, genome.ucsc.edu or www.ensembl.org) and represents the assembly of the genomes of a few young healthy donors. The inter-individual difference between all of the more than seven billion humans is only in the order of 0.1 %, while the difference to the closest relatives of man, the chimpanzees, is only some 4 %. Since the completion of the *Human Genome Project*, the whole genome sequence of thousands of humans has been determined (some of which in the context of the *1000 Genomes Project*, see Sect. 2.6).

It should be noted that the reference human genome as well as the results of all re-sequencing events still comprise a few problems:

1. The sequencing method is not perfect and there may be a technical error in about 1 of 100,000 nucleotides.
2. The genome contains a few hundred gaps. Most of these will be closed during the next years, but some 30 larger regions of complex and repetitive heterochromatin (primarily at the centromeres) may stay unsolved for a longer time.
3. Some 5 % of the human genome is highly repetitive, i.e. it contains very similar duplicated sequences that might cause problems in accurate sequence assembly.
4. CNVs are common variations of the human genome.

Box 2.4 Big Biology Projects

With a delay of some 20 years molecular biologists followed the example of physicists and realized that some of their research aims could only be reached by through multi-national collaborations of dozens to hundreds of research teams and institutions in so-called big biology projects. The *Human Genome Project* (www.genome.gov/10001772), which was launched in 1990 and finished in 2003, was the first example and together with follow-up studies it had

(continued)

Box 2.4 (continued)

a tremendous impact on the understanding of the architecture and function of the human genome. The *HapMap Project* (http://hapmap.ncbi.nlm.nih.gov) was one of these follow-ups benefitting from the advancing genotyping technologies. In parallel, improved next-generation sequencing methods allowed personal genome sequencing of both normal and cancer genomes. This made large-scale genome sequencing studies possible, such as the *1000 Genomes Project* (www.1000genomes.org) and the *International Cancer Genome Consortium* (https://icgc.org). Furthermore, the *Encyclopedia of DNA elements (ENCODE) Project* (www.genome.gov/encode) and *Functional Annotation of the Mammalian Genome (FANTOM5) Project* (http://fantom.gsc.riken.jp/5) focussed on the functional characterization of the human genome and will be discussed in more detail in Chaps. 4 and 5. Furthermore, the *International Human Epigenome Consortium* (www.ihec-epigenomes.org) that includes the *NIH Roadmap Epigenetics Project* (www.roadmapepigenomics.org) and *BLUEPRINT* (www.blueprint-epigenome.eu) plans to establish 1000 reference epigenomes from a diversity of primary human tissues and cell types.

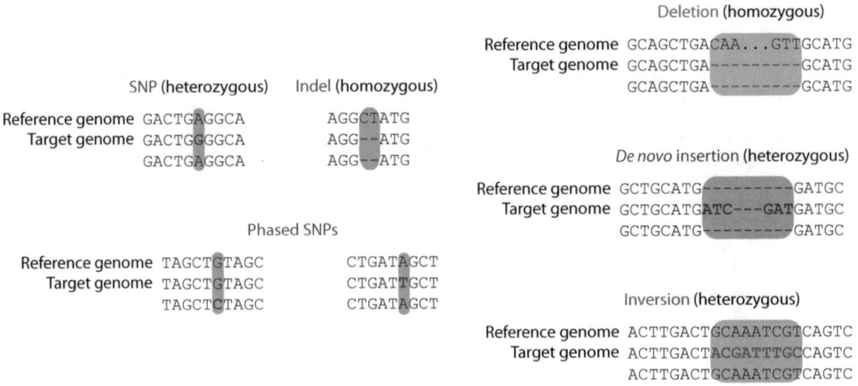

Fig. 2.3 Types of variations present in human genome sequences. The haploid reference genome is indicated at the top of each variant example, while the individual's diploid genome is shown below. The genetic variants can be either heterozygous or homozygous

within coding regions result in frameshift mutations, i.e. from the position of the mutation onwards the whole amino acid sequence is changed. In inversion variants the order of the bp is reversed, such as a 900 kb region of chromosome 17 that is in the reverse order in approximately 20 % of individuals with Northern European decent. Furthermore, CNVs consist of deletions or insertions of DNA streches in one genome compared to another. These variants can be heterozygous or homozygous. A pre-dominant class of insertions is that of mobile-element insertions, which are derived from ancient transpons. These DNA streches persist in the genome as SINEs (for example, *Alu* elements) and long interspersed nuclear elements (LINEs). More than 28,000 unique CNVs are known and some of them are quite common in human populations. The different types of human genetic variants are referred to as common (or polymorphisms), when they have a minor allele frequency (MAF) of at least 1 % in the studied population, or as rare, when they have a MAF of less than 1 %.

SNPs represent the most common class of genetic variations among human individuals. The human genome contains at least 11 million SNPs, of which approximately seven million show a MAF of more than 5 %, while the remaining are found in a frequency of 1–5 % in human populations. Information about them can be found, for example, in the *Single Nucleotide Polymorphism Database* (www.ncbi.nlm.nih.gov/SNP). The *1000 Genomes Project* (Sect. 2.6) indicated that there in addition is a huge number of rare and novel single nucleotide variants. Nevertheless, in any given individual, the majority of variants are common in the whole population. Moreover, the majority of the bp that differ between any pair of human individuals belong to the common variants. In addition, structural variants basically represent the remaining variants between individuals that are not SNPs. Since the detection of structural variants needs advanced technology, basically all initial associations between genome variations and complex traits, such as genome-wide association studies (GWASs, Sect. 2.5), were done only with SNPs. Nevertheless, for a given human individual, structural variants cover between 9 and 25 Mb of sequence, i.e. 0.5–1 % of the genome.

The average difference in nucleotide sequence of a pair of familial unrelated humans lies in the order of 1 in 1000. This proportion is low compared with other species and confirms the recent origin of *Homo sapiens sapiens* from a small founding population. The impact of SNPs on the coding sequence of the human genome is well established. Synonymous mutations do not alter the encoded protein, while non-synonymous mutations cause a change in the amino acid sequence (missense) or introduce a pre-mature stop codon (nonsense). Indels as well as CNVs in exonic sequences can result in either non-frameshift or frameshift mutations. Moreover, CNVs in intronic sequences may lead to alternative splicing. The impact of genetic variations in the non-coding region of the human genome will be discussed in Sect. 4.4.

2.4 The *HapMap Project* and Haplotype Blocks

The genetic approach of linkage mapping (Box 2.5) is used since decades and identified genes responsible for many inherited monogenetic disorders, such as the neurodegenerative Huntington's disease. However, these types of diseases represent only a relatively small fraction of all disorders. In contrast, most human diseases have a complex origin, i.e. they involve many gene loci in a complex interaction pattern. For these cases the genome-wide identification of SNPs via genotyping arrays as markers for disease-associated variants was found to be a more suitable approach. With an average SNP density of 1 in 1000, nucleotides studies on the genetic basis of phenotypic and trait diversity require the testing of millions of SNPs per individual and hundreds to thousands of subjects. For this purpose, large-scale studies (Box 2.4), such as the *Human Genome Diversity Project* and the *HapMap Project*, were launched. They used high-throughput SNP genotyping technologies, such as arrays with up to a million SNPs. *HapMap* started in 2002 with 270 samples from the three major human populations, which were 90 samples from Yoruba individuals living in Ibadan, Nigeria; each 45 samples from Han Chinese individuals living in Beijing, China and Japanese individuals living in Tokyo, Japan; and 90 samples from individuals with European ancestry living in Utah, USA. However, in its latest version, *HapMap 3*, the study extended to 1184 individuals from 11 global populations. The consortium performed genotyping for 1.6 million common SNPs and CNVs and used knowledge from linkage disequilibrium analysis of haplotype blocks (Box 2.5), in order to increase the efficiency of SNP-based mapping.

Box 2.5 A Population Genetics Glossary

Association studies: A method for localizing genes being responsible for specific diseases by comparing the DNA of a selected set of patients, who are believed to carry the same mutation/s because of their ancestral origin, with that of unrelated healthy controls from the same population.

Candidate gene studies: These types of hypothesis-driven studies aim to identify an association between genetic markers, such as SNPs, in genes or genomic regions that are pre-defined for the trait of interest. The candidates are selected based on i) biological evidence, ii) location in the genome or iii) evidence from gene disruptions, such as translocations or deletions.

GWAS: These studies use hypothesis-free methodologies, such as genome-wide SNP arrays, to identify genetic variants associated with traits of interest. The allele and/or genotype frequency of each SNP is tested for association with the investigated phenotypes. Genotyped SNPs are often not causal themselves, but are in linkage disequilibrium with the causal variants. GWAS of complex diseases require large sample numbers, in order to provide sufficient statistical power for the detection of associations of the typically small effect size of common variants.

Haplotype: A set of genetic markers that show complete or nearly complete linkage disequilibrium, i.e. the markers are inherited through generations without being changed by crossing-over or other recombination mechanisms.

Hardy-Weinberg equilibrium: A classical mathematical principle in population genetics used for testing random mating. It gives the expected frequencies of genotypes for a gene after one generation of random mating, if the parental allele frequencies are known.

(continued)

Box 2.5 (continued)

Linkage analysis: This approach attempts to identify regions in chromosomes that co-segregate with the trait of interest in related individuals. Due to the inverse relationship between the frequency of recombination between two loci on a chromosome and the physical distance between them, a co-inheritance of genetic markers with a disease phenotype suggest that they are "linked", i.e. they are located in close proximity. Linkage analyses can be performed with microsatellites, i.e. polymorphic markers, or SNPs.

Linkage disequilibrium: An association between two alleles that are located so close to each other on the genome that they are inherited together more frequently than expected by chance.

Minor-allele frequency (MAF): The proportion of the less common of two alleles in a population, ranging from 1 to 50 %.

Haplotypes are stretches of genomic DNA of typically 10–100 kb in length that are inherited from generation to generation in blocks. In this way, a few "tag" SNPs can serve as representatives for the great majority of SNP variation within each block (see example in Fig. 2.4). The borders of haplotype blocks represent recombination events during meiosis that has happened in ancestors of present day humans. Since African populations have existed at least twice as long as European and Asian populations, their haplotype blocks are shorter, i.e. they had more time to decay because of the accumulation of recombination events in a higher number of generations (Fig. 2.5). In contrast, all non-African humans derived from a small population of eastern African origin, i.e. they went through a demographical bottleneck that is clearly visible in the genomes of modern humans. The *HapMap Project* used haplotype blocks of sufficient stability and length, in order to reduce the number of SNPs that would need to be genotyped. For example, if haplotypes contained 20 SNPs on average, only one of these SNPs would be needed to tag a haplotype. Table 2.1 provides a typical *HapMap Project* result at the example of the SNP rs1544410 within the vitamin D receptor (*VDR*) gene, which plays a critical role in predicting bone mineral density (BMD) in human females.

2.5 Genome-wide Association Studies

The technology development in context of the *HapMap Project*, i.e. array-based genotyping and analysis of at least 100,000 tag SNPs in parallel, dramatically changed the field of the human disease genetics. It shifted from candidate-gene association and linkage studies to GWASs. The latter employ an "agnostic" approach in the search for unknown disease variants, for which the ability to interrogate a large number of SNPs covering the entire human genome is essential, i.e. hundreds

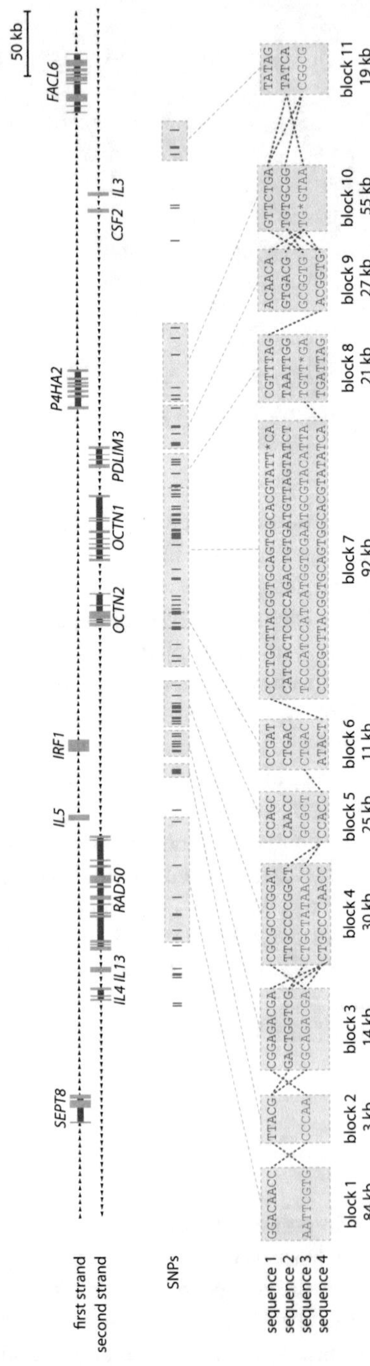

Fig. 2.4 Block-like haplotype diversity and risk for disease. A 500 kb region of human chromosome 5q31 that contains a genetic risk factor for Crohn's disease was used as an example. The genomic region is subdivided into 11 common haplotype blocks of 11–92 kb in size, for which tag SNPs are indicated

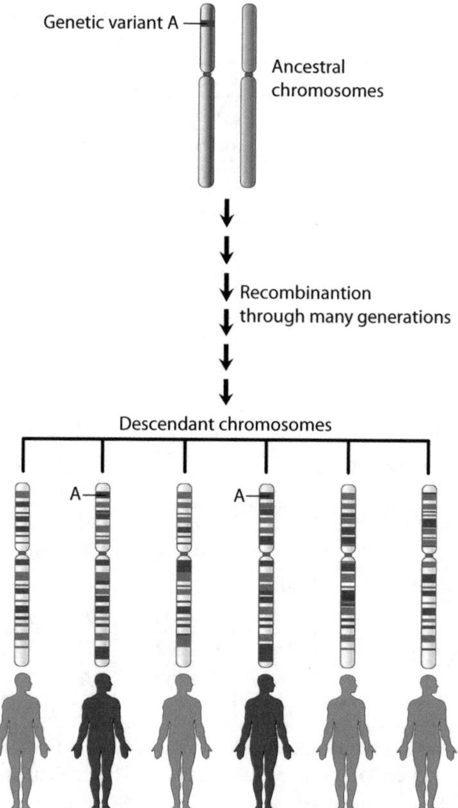

Fig. 2.5 The origin of haplotypes. Two ancestral example chromosomes get scrambled through meiosis-related recombination over many generations, in order to yield different descendant chromosomes. For example, after 30,000 years a typical chromosome will have undergone more than one crossover per 100 kb. In the case of a genetic variant (marked by the A) on one ancestral chromosome the risk of a particular disease increases. Thus, the two individuals in the current generation who inherited that region of the ancestral chromosome (referred to as a haplotype block) will be at increased risk. Within the same haplotype block, which carries the disease-causing variant, there are many SNPs that can be used to identify the location of the variant

of thousands of SNPs are tested for association with a disease in hundreds or thousands of human individuals. 10 years of intensive GWAS research resulted in more than 2400 publications reporting some 16,600 SNPs (as of April 2016) being statistically robustly associated with one out of more than 500 complex diseases and traits (see the database *Catalog of Published Genome-Wide Association Studies*, www.ebi.ac.uk/gwas). A number of associations that are identified in one population are often not found in members of other populations. Therefore, the most reliable evidence of a true genetic association is its replication in multiple populations. However, most published studies have focused primarily on populations of European ancestry.

Table 2.1 Multiple studies examining bone mineral density (BMD) in women have associated the A allele with increased risk of low BMD causing osteoporosis

Handle-polulation Id	2n	Alelle freq	Genotype freq	Hardy-Weinberg
TSC-CSHL-SC 12 A	22	**A** 0.409	**G/G** 0.182	Chi Square 5.272
		G 0.591	**A/G** 0.818	
TSC-CSHL-SC 12 AA	14	**G** 0.5	**A/G** 1.000	Chi Square 7
		A 0.5		
TSC-CSHL-SC 12 C	14	**G** 0.5	**A/A** 0.286	Chi Square 0.143
		A 0.5	**G/G** 0.286	
			A/G 0.429	
TSC-CSHL-SC 95 C	88	**A** 0.432	**A/A** 0.205	Chi Square 0.239
		G 0.568	**G/G** 0.341	
			A/G 0.455	
CSHL-HAPMAP-HapMap-CEU	120	**A** 0.442	**A/A** 0.233	Chi Square 0.144
US (residents with ancestry from Northern and Western Europe)		**G** 0.**558**	**G/G** 0.35	
			A/G 0.417	
CSHL-HAPMAP-HapMap-HCB	90	**A** 0.022	**A/G** 0.044	Chi Square 0.024
China		**G** 0.978	**G/G** 0.**956**	
CSHL-HAPMAP-HapMap-JP**T**	88	**A** 0.136	**A/G** 0.273	Chi Square 1.102
Japan		**G** 0.864	**G/G** 0.727	
CSHL-HAPMAP-HapMap-YR1	118	**A** 0.297	**A/A** 0.136	Chi Square 3.069
Nigeria (Yoruba)		**G** 0.703	**A/G** 0.322	
			G/G 0.542	

The SNP rs1544410, also known as the *BsmI* polymorphism, is located within the *VDR* gene. Multiple studies examining BMD in women have associated the A allele with increased risk of low bone minaral density (BMD) causing osteoporosis. The allele frequency, genotype frequency and the Hardy-Weinberg equilibrium statistics are indicated for different *HapMap* populations

With the decreasing cost of the SNP arrays, it became feasible to genotype thousands of samples in GWASs. Despite some notable successes in revealing numerous novel SNPs and loci associated with complex phenotypes, most of the complex, polygenic traits have not had much more than 10 % of their heritability explained by the common variants assessed by GWAS. The "missing" or unsolved heritability does not allow assigning an individual with any reliable estimation about his/her risk for a particular disease. The only well-known exceptions are age-related macular degeneration and type 1 diabetes (T1D), for which the combinations of common and rare variants can provide a quantifiable risk profile. GWASs with 2000–5000 individuals confidently identify common variants with effect sizes, referred to as odds ratios (ORs), of 1.5 or greater. This makes it unlikely that further common SNPs with moderate or even large ORs in complex traits will be discovered in future. Limited statistical power to detect small gene-gene and gene-environment interactions requires increasing sample sizes, which may be achieved by

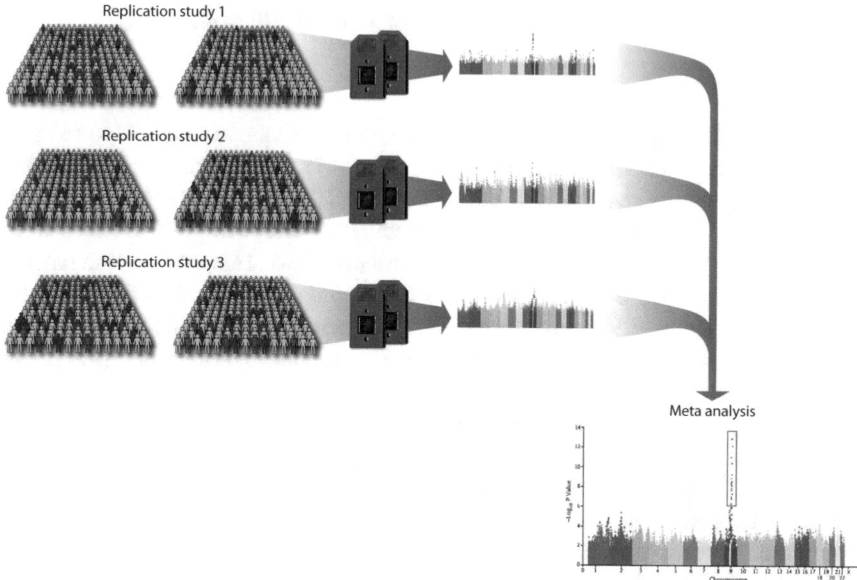

Fig. 2.6 GWAS meta-analysis. The statistical power of GWASs for the detection of associations can be increased by a meta-analysis computing the effects of multiple studies. Here a fictive example is shown, in which the results of three studies, which each did not reach significant associations, are combined in a meta-analysis now detecting a strong, significant association on chromosome 9

pooling several GWASs through meta-analysis (Fig. 2.6). For example, sample sizes of 60,000 subjects are necessary to provide sufficient power to identify the majority of variants with ORs of 1.1 (i.e. a 10 % increased risk for the tested disease). Nevertheless, the majority of the missing heritability may be due to rare variants with high ORs, which are poorly captured by standard GWASs, or due to epigenetic effects that cannot be detected by present genotyping arrays (Chap. 5).

GWASs indicate SNPs that are in high linkage disequilibrium to the genomic variants that cause the disease, i.e. often they are on the same haplotype block. However, this implies that most disease-associated SNPs are not functionally relevant to mechanisms of the respective disease. Extensive sequencing of an associated region may identify additional, previously unknown, rare variants with a possible biologic role. This is one of the goals of the *1000 Genomes Project* (Sect. 2.6). However, protein-coding regions carry only 12 % of all SNPs that are associated with traits, i.e. only in these cases a straightforward mechanistic explanation of the impact of the genetic variation is possible. This implies that many of the remaining 88 % variations may affect the genomic binding sites of transcription factors, i.e. they are regulatory SNPs (Sect. 4.4)

2.6 Whole Genome Sequencing and the *1000 Genomes Project*

Important technological advances in high-throughput sequencing have led to rapid decrease in the costs of DNA sequencing (Fig. 2.7). As a result, whole genome sequencing becomes an affordable tool in understanding the genomic basis of health and disorders. At present, already huge amounts of data have been obtained from whole genomes of both healthy and diseased individuals. This information has not only helped in disease stratification and in the identification of their molecular mechanisms, but also is transforming the perspective of future health care from disease diagnosis and treatment to personalized health monitoring and preventive medicine.

"Whole genome sequencing" results in the identification of the complete set of genetic variants of a given human individual (Fig. 2.8). As a natural extension of the *HapMap Project*, the *1000 Genomes Project* was initiated in 2008 with the aim to sequence the genomes of at least 1000 individuals from different populations around the world. *HapMap* populations are included in the project (Table 2.2), and in total 2504 genomes from 26 populations covering all five continents are investigated. In

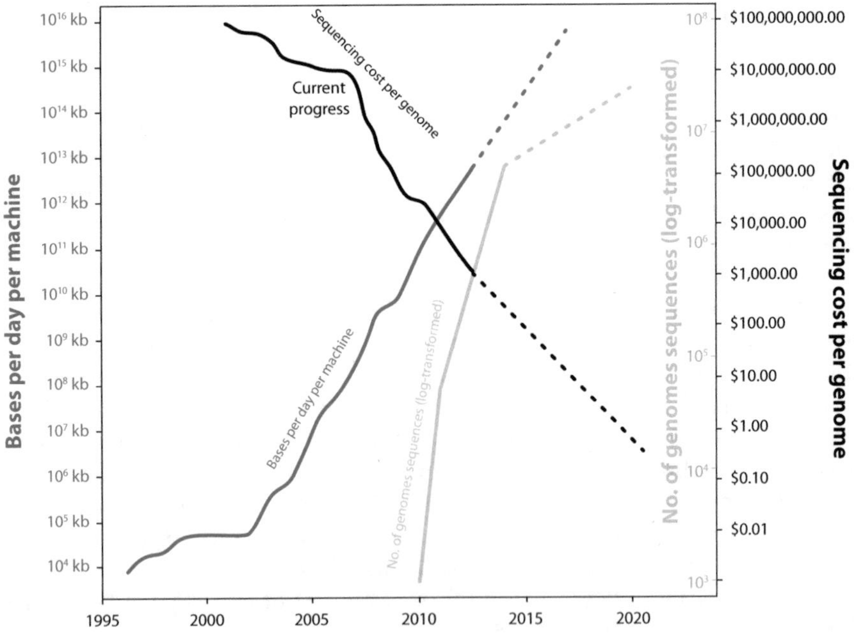

Fig. 2.7 Whole genome sequencing costs. After many years of decline, the cost of sequencing a genome had leveled off, but may dive again (*dashed line*) when further technological advances will become effective. For more details see www.genome.gov/sequencingcosts

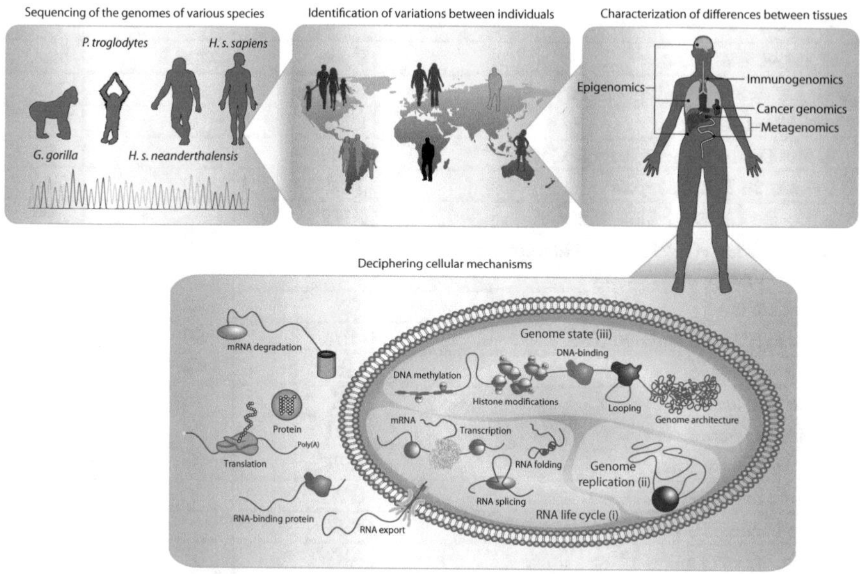

Fig. 2.8 Road map of sequencing science. The *Human Genome Project* created a reference genome and now also the genomes of all other primate species are known including some extinct human species (*top left*). Whole genome sequencing of several thousand human individuals is performed in large consortia, such as the *1000 Genomes Project* (*top center*). Moreover, the genetic and epigenetic differences between tissues and cell types of the same human individual are collected in cancer genomics and epigenomics projects, such as the *Cancer Genome Consortium* and the *Human Microbiome Project* (*top right*). These sequencing provide a window into the illustrated cellular processes (*bottom*)

total, the project describes 88 million genome variants, of which 84.7 million are SNPs, 3.6 million short indels and 60,000 structural variants. As expected based on the out-of-Africa model of human origins, individuals from African ancestry populations show most variant sites. A typical human genome carries 200,000 variants, most of which are common and only 20 % are rare (MAF < 0.5 %). In average, a typical genome contains some 150 variants resulting in protein truncation, 10,000 changing amino acids and 500,000 affecting transcription factor binding sites. The data demonstrate that each human individual seems to differ from the current reference genome by putative loss-of-function mutations in 150 or more genes. Moreover, the results imply that rare variants in individuals with a particular disease should be interpreted within the context of their geographic and/or ancestry-based genetic background.

The rapid maturation of next-generation sequencing technologies led to the exponential development of methods for nearly all aspects of cellular processes, such as ChIP-seq, RNA-seq and FAIRE-seq, i.e. sequencing emerged to a readout for their detailed and comprehensive analysis (Fig. 2.8). Fast and inexpensive next-generation sequencing has enabled a number of other large-scale studies, such as

Table 2.2 Populations of the *1000 Genomes Project*

Full population name	Code	Phase 1	Phase 3
Han Chinese in Beijing, China	**CHB**	**97**	**103**
South Han Chinese	CHS	100	105
Chinese Dai in Xishuangbanna, China	CDX		93
Japanese in Tokyo, Japan	**JPT**	**89**	**104**
Kinh in Ho Chi Minh City, Vietnam	KHV		99
Utah residents (CEPH) with Northern and Western European ancestry	**CEU**	**85**	**99**
Toscani in Italy	TSI	98	107
British in England and Scotland	GBR	89	91
Finnish in Finland	FIN	93	99
Iberian Populations in Spain	IBS	14	107
Yoruba in Ibadan, Nigeria	**YRI**	**88**	**108**
Luhya in Webuye, Kenya	LWK	97	99
Gambian in Western Division, Mandinka	GWD		113
Mende in Sierra Leone	MSL		85
African Ancestry in Southwest US	ASW	61	61
Esan in Nigeria	ESN		99
African Caribbean in Barbados	ACB		96
Mexican Ancestry in Los Angeles, CA, USA	MXL	66	64
Colombian in Medellin, Colombia	CLM	60	94
Peruvian in Lima, Peru	PEL		85
Puerto Rican in Puerto Rico	PUR	55	104
Bengal in Bangladesh	BRB		86
Gujarati Indians in Houston, TX, USA	GIH		103
Indian Telugu in the UK	ITU		102
Sri Lankan Tamil in the UK	STU		102
Punjabi in Lahore, Pakistan	PJL		96
TOTALS		1092	**2504**

The human populations that are included in the *1000 Genomes Project* are listed. The numbers of individuals that were investigated in phase 1 of the project and the final deep-coverage sequencing of 2504 subjects (phase 3) are indicated. Original *HapMap* populations are highlighted in *bold*

the *ENCODE Project* and the *FANTOM5 Project*, which together produced the presently most comprehensive map of functional genomic elements within the human genome. In Chap. 4 we will discuss further applications of these projects including the identification and characterization of regulatory SNPs (Sect. 4.4) and the integrative personal omics profile (iPOP) of one human individual (Sect. 4.6).

Future View

Within only a few years next-generation sequencing technologies significantly improved our knowledge of human genetic diseases. It is expected that this trend will continue. Although data privacy of human subjects is a key issue, it will be increasingly difficult to control and may deter individuals to participate in genetic studies. However, many healthy as well as diseased people support an open data concept and have no problem in making their genomic information public. Interestingly, each human individual seems to be heterozygous for 50–100 genetic variants that may cause inherited disorders in homozygous offspring. This will provide a large demand and challenge for genetic counselling based on whole-genome sequencing. Moreover, gene-environment interactions provided by lifestyle factors, as the personal choice of food, will create an additional level of complexity. Within the next few years a number of next-generation sequencing applications will be incorporated into clinical diagnostics, but it is yet unclear, how this will be financed. Nevertheless, further developments of next-generation sequencing will stay a driving force in basic biomedical research, including nutrigenomics.

Key Concepts

- The anatomically modern human developed some 200,000 years ago in Africa and started some 60–80,000 years ago to migrate towards Eurasia.
- Some 10,000 years ago, humans started to become farmers implying significant lifestyle changes, such as the use of cereals and milk as new type of diet, being followed from immunological challenge via enhanced rates of microbe infections.
- In the context of an inactive lifestyle combined with energy-dense foods, gene variants that had been initially evolved for an efficient energy storage and physical mobility increase the risk for chronic diseases, such as T2D or CVD.
- Hundreds of complex phenotypic traits determine the risk to develop complex diseases each being based on dozens to hundreds of gene variants and environmental influences.
- Every human individual carries, on average, approximately four million genetic variants covering about 12 Mb of sequence, i.e. 0.3 % of the whole genome.
- Human genetic variants are referred to as common, when they have a MAF of at least 1 % in the studied population, or as rare, when they have a MAF of less than 1 %.
- The average difference in the nucleotide sequence of two unrelated humans is in the order of 1 in 1000.
- The *HapMap Project* was launched in order to map genetic variants of the main human populations via genome-wide genotyping.
- Despite some notable successes in revealing numerous novel SNPs and loci associated with complex phenotypes, all GWAS-SNPs collectively account for only some 10 % of the heritability of complex diseases.
- GWASs indicate SNPs that are in high linkage disequilibrium to the genomic variants that cause the disease, but most disease-associated SNPs are not functionally relevant to mechanisms of the respective disease.
- Whole genome sequencing results in the identification of the complete set of genetic variants of a given human individual, such as the rare SNPs with large effect sizes that significantly contribute to complex traits and diseases.
- The *1000 Genomes Project* is the natural extension of the *HapMap Project* and uses whole genome sequencing of 2504 genomes from populations covering all five continents.

Additional Reading

1000 Genomes Project Consortium, Auton A, Brooks LD, Durbin RM, Garrison EP, Kang HM, Korbel JO, Marchini JL, McCarthy S, McVean GA, Abecasis GR (2015) A global reference for human genetic variation. Nature 526:68–74

Abecasis GR, Auton A, Brooks LD, DePristo MA, Durbin RM, Handsaker RE, Kang HM, Marth GT, McVean GA (2012) An integrated map of genetic variation from 1,092 human genomes. Nature 491:56–65

Altshuler DM, Gibbs RA, Peltonen L, Altshuler DM, Gibbs RA, Peltonen L, Dermitzakis E, Schaffner SF, Yu F et al (2010) Integrating common and rare genetic variation in diverse human populations. Nature 467:52–58

Haraksingh RR, Snyder MP (2013) Impacts of variation in the human genome on gene regulation. J Mol Biol 425:3970–3977

Manolio TA (2010) Genomewide association studies and assessment of the risk of disease. N Engl J Med 363:166–176

Pääbo S (2014) The human condition – a molecular approach. Cell 157:216–226

Shendure J, Lieberman Aiden E (2012) The expanding scope of DNA sequencing. Nat Biotechnol 30:1084–1094

Sudmant PH, Rausch T, Gardner EJ, Handsaker RE, Abyzov A, Huddleston J, Zhang Y, Ye K, Jun G, Hsi-Yang Fritz M, Konkel MK, Malhotra A, Stütz AM, Shi X, Paolo Casale F, Chen J, Hormozdiari F, Dayama G, Chen K, Malig M, Chaisson MJ, Walter K, Meiers S, Kashin S, Garrison E, Auton A, Lam HY, Jasmine Mu X, Alkan C, Antaki D, Bae T, Cerveira E, Chines P, Chong Z, Clarke L, Dal E, Ding L, Emery S, Fan X, Gujral M, Kahveci F, Kidd JM, Kong Y, Lameijer EW, McCarthy S, Flicek P, Gibbs RA, Marth G, Mason CE, Menelaou A, Muzny DM, Nelson BJ, Noor A, Parrish NF, Pendleton M, Quitadamo A, Raeder B, Schadt EE, Romanovitch M, Schlattl A, Sebra R, Shabalin AA, Untergasser A, Walker JA, Wang M, Yu F, Zhang C, Zhang J, Zheng-Bradley X, Zhou W, Zichner T, Sebat J, Batzer MA, McCarroll SA, 1000 Genomes Project Consortium, Mills RE, Gerstein MB, Bashir A, Stegle O, Devine SE, Lee C, Eichler EE, Korbel JO (2015) An integrated map of structural variation in 2,504 human genomes. Nature 526:75–81

Veeramah KR, Hammer MF (2014) The impact of whole-genome sequencing on the reconstruction of human population history. Nat Rev Genet 15:149–162

Part II
Molecular Genetic Basis

Chapter 3
Sensing Nutrition

Abstract Sensing fluctuations in nutrient levels is essential for life. Therefore, mammals have evolved distinct mechanisms to sense the abundance of lipids, amino acids and glucose. Nutrient-sensing pathways stimulate anabolism and storage, when food is abundant, whereas scarcity induces homeostatic processes, such as the mobilization of internal energy stores. Transcription factors, in particular members of the nuclear receptor superfamily, are the proteins through which nutrients can directly influence gene expression. Many nuclear receptors bind micro- and macronutrients or their metabolites, such as fatty acids to PPARs, oxysterols to liver X receptors (LXRs), bile acids to farnesoid X receptor (FXR) and vitamin D metabolites to VDR. In particular in metabolic organs, such as liver, pancreas, intestine and adipose tissue, nuclear receptors respond to nutrient changes and specifically activate hundreds of their target genes. Moreover, also the immune system is triggered in its inflammatory and antigen response by nuclear receptors and their ligands. In addition, nuclear receptors belong to those transcription factors that play a central role in managing the molecular clock both in the CNS as well as in metabolic organs.

In this chapter, we will discuss nutrient sensing mechanisms via membrane receptors, metabolic enzymes, regulatory kinases and transcription factors. We will get insight into the central role of nuclear receptors in the translation of nutrient fluctuations into responses of the genome. We will use the nuclear receptors PPARs, LXR and FXR as examples, in order to understand their important function in controlling fatty acid, cholesterol and bile acids transport and metabolism. Furthermore, we will demonstrate the impact of vitamin D and its receptor VDR on the function of both the innate and the adaptive immune system. Finally, we will integrate the knowledge on nutrient sensing with information on circardian processes controlling anabolism and catabolism.

Keywords Lipid sensing • Amino acid sensing • Glucose sensing • Nuclear receptors • Gene regulation • Target genes • Lipid metabolism • PPARs • LXRs • FXR • VDR • Innate immunity • Adaptive immunity • Molecular clock • REV-ERB • ROR

C. Carlberg et al., *Nutrigenomics*, DOI 10.1007/978-3-319-30415-1_3

3.1 Nutrient-Sensing Mechanisms

Periodical scarcity of nutrients was a strong evolutionary pressure to select efficient mechanisms of nutrient sensing. This may be either the direct binding of the molecule to its sensor or an indirect mechanism that is based on the detection of a surrogate molecule that reflects the nutrient's abundance. The respective sensor is a protein that binds the nutrient with an affinity in the order of the fluctuations its physiological concentrations. The sensing of nutrients may then trigger the release of hormones acting as long-range signals, in order to achieve a coordinated response of the whole organism.

Due to their non-polar nature, i.e. their insolubility in aqueous solutions, lipids are rarely found free in soluble form but are either contained in lipoproteins (Sect. 11.3) or bound in the serum by albumin. G protein-coupled receptors (GPRs), such as GPR40 and GPR120, bind long-chain unsaturated fatty acids, for example, in the membrane of β cells and enhance in these cells glucose-triggered insulin release (Fig. 3.1a). In WAT, the binding of glucose to GRP120 induces a signal transduction cascade involving the kinases PI3K and AKT resulting in glucose uptake. In the intestinal lumen, fatty acids are bound by the scavenger receptor CD36 molecule (CD36) that initiates their uptake (Fig. 3.1b). Adequate sensing of internal cholesterol levels is important, in order to avoid, in case of abundant external supply, the activation of the energetically demanding cholesterol biosynthetic pathway and to prevent toxic levels of free cholesterol in the cell. Cholesterol binds to the protein SREBF chaperone (SCAP) (Fig. 3.1c). In case of high cholesterol levels, SCAP increases its affinity for insulin-induced gene (INSIG) protein that anchors SCAP and the transcription factor sterol regulatory element-binding transcription factor 1 (SREBF1) within the ER membrane. When cholesterol levels are low, the SCAP-SREBF complex dissociates from INSIG and shuttles to the Golgi apparatus, where SREBF is released, translocates to the nucleus and activates genes involved in lipid anabolism, i.e. cholesterol biosynthesis and lipogenesis (Fig. 3.1d). In addition, at low cholesterol levels the enzyme HMG-CoA reductase (HMGCR), which is also located in the ER membrane, catalyzes a rate-limiting step of the cholesterol *de novo* synthesis (Fig. 3.1e). In contrast, high levels of intermediates in the cholesterol biosynthesis pathway, such as lanosterol, trigger the binding of HMGCR to INSIG, which leads to the ubiquitin-mediated degradation of the enzyme (Fig. 3.1fc). LXRs are activated by elevated cholesterol levels (Sect. 3.4), i.e. the pathways of SREBF and LXRs work in a reciprocal fashion, in order to maintain cellular and systemic cholesterol homeostasis.

In case of shortage in amino acids, cellular proteins are used as a reservoir and degraded via the proteasome or by autophagy. Autophagy is a basic catabolic mechanism that involves the degradation of unnecessary or dysfunctional cellular components through the actions of lysosomes. This process promotes survival during starvation by maintaining cellular energy levels. Amino acids are catabolized for the production of glucose and ketone bodies, i.e. essential energy sources for the brain,

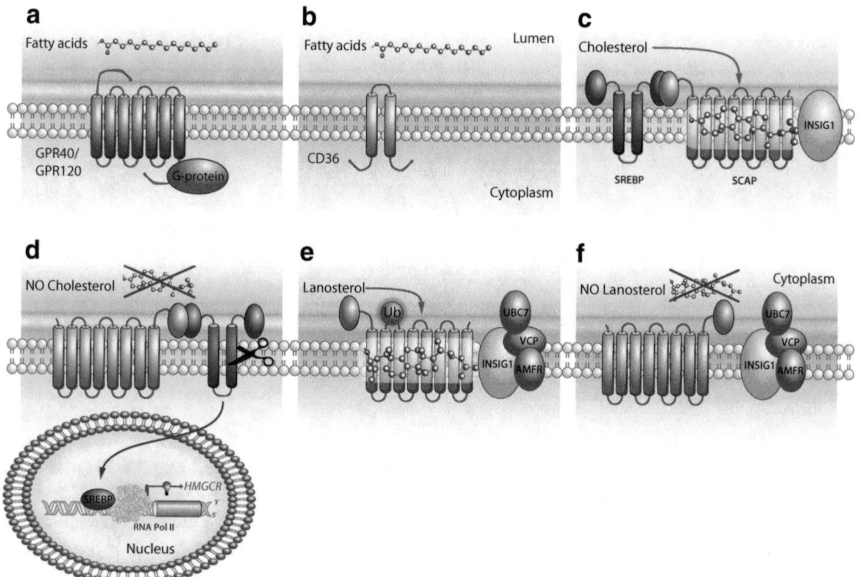

Fig. 3.1 Lipid-sensing mechanisms. (**a**). Fatty acid detection mechanisms by GPR40 and GPR120 (*left*) and CD36 (*right*). In enteroendocrine cells the binding of lipids to GPRs leads to the release of incretins, such as glucagon-like peptide 1 (GLP1), in taste buds it triggers the release of neurotransmitters and in WAT it promotes glucose uptake. (**b**). Binding of fatty acids to CD36 triggers in taste buds calcium release from the ER and neurotransmission, while in enterocytes it promotes fatty acid uptake. (**c**). In the presence of cholesterol, the SCAP-SREBF complex binds INSIG proteins and remains anchored to the ER. (**d**). In the absence of cholesterol SCAP-SREBF does not bind INSIG but moves to the Golgi, where the cytoplasmic tail of SREBF is cleaved. This stimulates the transcription of genes involved in cholesterol synthesis. (**e**). The ER-embedded enzyme HMGCR catalyzes a rate-limiting step in cholesterol synthesis and is expressed at low cholesterol levels. (**f**). When intermediates of the cholesterol biosynthetic pathway, such as lanosterol, are abundant, HMGCR interacts with INSIG proteins leading to HMGCR ubiquitination and degradation

during periods of prolonged starvation and hypoglycemia. The target of rapamycin (TOR) component mTORC1 is able to sense amino acids (Sect. 6.1). Ras-related GTP binding (RRAG) proteins recruit mTORC1 to the outer lysosomal surface, in order to activate it via the protein Ras homolog enriched in brain (RHEB). The lysosome is an organelle, in which amino acids and other nutrients are scavenged from cellular components through autophagy. Therefore, high levels of amino acids within the lysosome reflect abundance of amino acids within the cell.

The plasma membrane protein GLUT2 (encoded by the gene *SLC2A2*) is a transporter with a rather low affinity for glucose (Fig. 3.2a). In contrast to the other high-affinity glucose transporters, GLUT2 acts as a true sensor for glucose, since it is only active at high but not at low physiologic glucose concentrations. Therefore, GLUT2 has a central role in directing the handling of glucose after

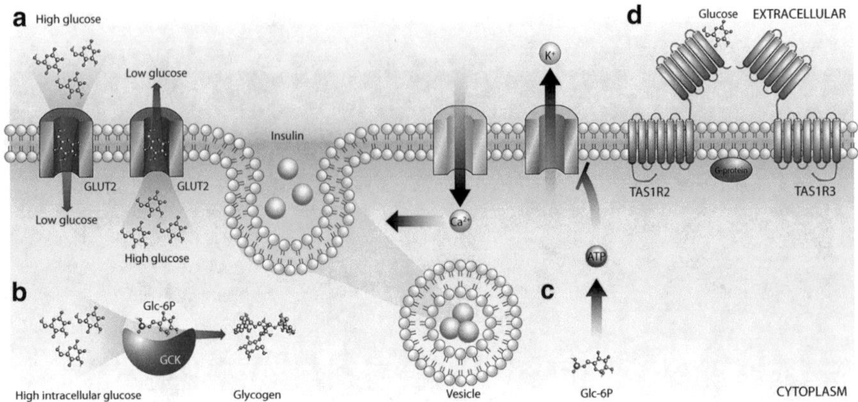

Fig. 3.2 Glucose-sensing mechanisms. (**a**). Due to its low affinity for glucose the transporter GLUT2 imports glucose only, when it has high concentrations (*right*) and exports glucose from hepatocytes into the circulation at hypoglycemic conditions (*left*). (**b**). The enzyme GCK has low affinity for glucose, i.e. only at high glucose concentrations it produces glucose-6-phosphate for the use in glycolysis or glycogen synthesis. (**c**). The release of insulin from β cells is a multi-step process that involves the phosphorylation of glucose by GCK, subsequent ATP production and ATP-mediated blocking of potassium channels. A resulting calcium influx facilitates the release of insulin from vesicles into the bloodstream. (**d**). The heterodimeric oral taste receptors TAS1R2-TAS1R3 bind only high concentrations of glucose, sucrose, fructose and artificial sweeteners and trigger signal transduction through G proteins

feeding. At periods of low glycemia, hepatic gluconeogenesis increases the glucose levels within liver cells, and GLUT2 exports glucose to the circulation. The intra-cellular sensing of glucose is mediated by the enzyme glucokinase (GCK) that catalyzes the first step in the storage and consumption of glucose, i.e. glycogen synthesis and glycolysis. GCK has a significantly lower affinity for glucose than the other hexokinases, i.e. it is only active at high glucose concentrations, and therefore functions, similar to GLUT2, as a glucose sensor (Fig. 3.2b).

At low glucose levels this property allows GCK in collaboration with GLUT2 to export unphosphorylated glucose from the liver to the brain and the muscles. β cells of the pancreas sense systemic glucose levels. Glucose is imported into β cells by GLUT2 and phosphorylated by GCK leading to an increased adenosine triphosphate (ATP)/adenosine diphosphate (ADP) ratio. This depolarizes the membrane via closing of potassium channels at the plasma membrane, leads to a transient increase of intra-cellular Ca^{2+}, stimulates the fusion of insulin-loaded vesicles with the plasma membrane and allows their release into systemic circulation (Fig. 3.2c). The detection of high energetic food in taste buds is mediated by the protein taste receptor, type 1, member 2 (TAS1R2), in complex with TAS1R3 (Fig. 3.2d). Millimolar concentrations of glucose, fructose or sucrose but also artificial sweeteners, such as saccharine, cyclamate and aspartame, activate the T1R2-T1R3 receptor that results in sensing "sweet".

3.2 Nutrient Sensing via Nuclear Receptors

Most extra-cellular signaling molecules, such as growth factors and cytokines, are hydrophilic and cannot pass cellular membranes, i.e. they need to interact with membrane receptors, in order to activate a signal transduction cascade that eventually leads to the activation of a transcription factor. Therefore, transcription factors can be considered as sensors of many types of perturbations of a cell. In contrast, in case of lipophilic signaling molecules, such as steroid hormones, the signal transduction process is more straightforward, since these compounds can pass cellular membranes and bind directly to a transcription factor that is often already located in the nucleus (Fig. 3.3).

The nuclear receptor superfamily has 48 members in humans and comprises special types of transcription factors that are able to bind and to be activated by small lipophilic molecules called ligands. Many of these nuclear receptor ligands are

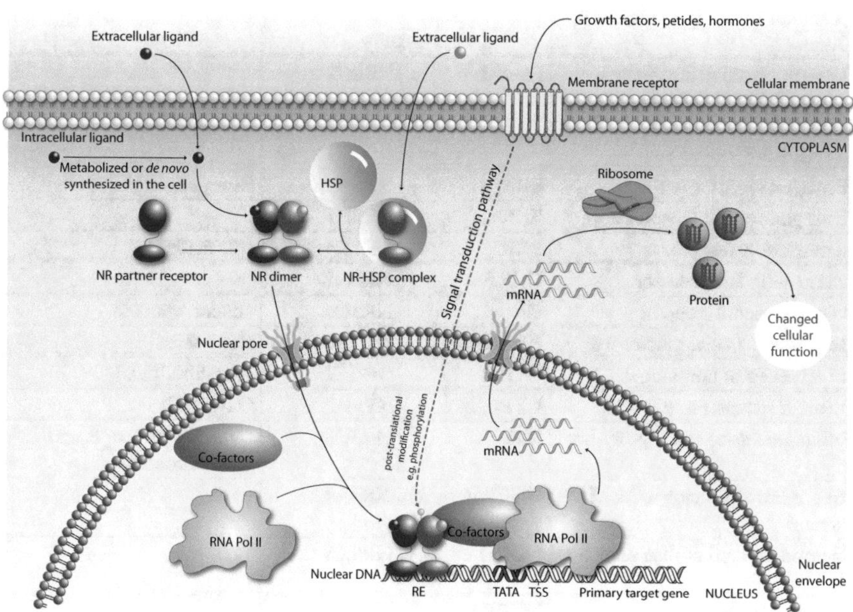

Fig. 3.3 Principles of nuclear receptor signaling. Some nuclear receptors, such as glucocorticoid receptor (GR) and androgen receptor (AR), reside in the cytoplasm in a complex with chaperone proteins, such as heat-shock proteins (HSPs), but most nuclear receptors are already located in the nucleus, where they are activated through the binding of their specific lipophilic ligand. The ligand is either of extra-cellular origin and has passed cellular membranes or is a metabolite that was synthesized inside the cell. After ligand binding, cytoplasmic nuclear receptors dissociate from their chaperones and translocate to the nucleus, where they bind, like the other members of the superfamily, to their specific genomic binding sites (REs) in the vicinity of transcription start site (TSS) regions of their primary target genes. Ligand-activated nuclear receptors interact with nuclear co-factors that build a bridge to the basal transcriptional machinery with Pol II in its core. This then leads to changes in the mRNA and protein expression of the target genes

micro- and macronutrients or their metabolites. These include the vitamin A derivative retinoic acid (activating retinoic acid receptor (RARα, β, γ)), fatty acids (PPARα, δ, γ), 1,25-dihydroxyvitamin D_3 (1,25$(OH)_2D_3$, VDR), oxysterols (LXRα, β), bile acids (FXR) and other hydrophobic food ingredients (constitutively androstane receptor (CAR) and pregnane X receptor (PXR)) (Table 3.1 and Fig. 3.4). The affinity of these nuclear receptors for their respective ligands varies between 0.1 nM for VDR and up to mM for PPARs and reflects the physiological concentrations of the respective molecules. Therefore, they represent true nuclear sensors for these

Table 3.1 Human nuclear receptor nomenclature and ligands

Common name	Common abbreviation	Unified nomenclature	Ligands
Androgen receptor	AR	NR3C4	Androgens
Constitutive androstane receptor	CAR	NR1I3	Xenobiotics
Chicken ovalbumin upstream promoter-transcription factor α, β, γ	COUP-TFα, β, γ	NR2F1, 2, 6	
Dosage-sensitive sex reversal-adrenal hypoplasia congenital critical region on the X chromosome, gene 1	DAX-1	NR0B1	
Estrogen receptor α, β	ERα, β	NR3A1, 2	Estrogens
Estrogen-related receptor α, β, γ	ERRα, β, γ	NR3B1, 2, 3	
Farnesoid X receptor	FXR	NR1H4	Bile acids
Germ cell nuclear factor	GCNF	NR6A1	
Glucocorticoid receptor	GR	NR3CI	Glucocorticoids
Hepatocyte nuclear factor 4 α, γ	HNF4α, γ	NR2A1	[fatty acids]
Liver receptor homolog-1	LRH-1	NR5A2	[phospholipids]
Liver X receptor α, β	LXRα, β	NR1H3, 2	Oxysterols
Mineralocorticoid receptor	MR	NR3C2	Mineralocorticoids and glucocorticoids
Nerve-growth-factor-induced gene B	NGF1-B	NR4A1	
Neuron-derived orphan receptor 1	NOR-1	NR4A3	
Nur-related factor 1	NURR1	NR4A2	
Photoreceptor-cell-specific nuclear receptor	PNR	NR2E3	
Peroxisome proliferator-activated receptor α, δ, γ	PPARα, δ, γ	NR1C1, 2, 3	Fatty acids
Progesterone receptor	PR	NR3C3	Progesterone
Pregnane X receptor	PXR	NR1I2	Endobiotics and xenobiotics
Retinoic acid receptor α, β, γ	RARα, β, γ	NR1B1, 2, 3	Retinoic acids
Reverse-Erb α, β	REV-ERBα, β	NR1D1, 2	[heme]
RAR-related orphan receptor α, β, γ	RORα, β, γ	NR1F1, 2, 3	[sterols]

(continued)

Table 3.1 (continued)

Common name	Common abbreviation	Unified nomenclature	Ligands
Retinoid X receptor α, β, γ	RXRα, β, γ	NR2B1, 2, 3	9-*cis* retinoic acid
Steroidogenic factor 1	SF-1	NR5A1	[phospholipids]
Short heterodimeric partner	SHP	NR0B2	
Tailless homolog orphan receptor	TLX	NR2E1	
Testicular orphan receptor 2, 4	TR2, 4	NR2C1, 2	
Thyroid hormone receptor α, β	TRα, β	NR1A1, 2	Thyroid hormones
Vitamin D receptor	VDR	NR1I1	1α,25(OH)$_2$D$_3$ and lithocholic acid

Atypical ligands that are bound constitutively to their receptors are indicated by *brackets*

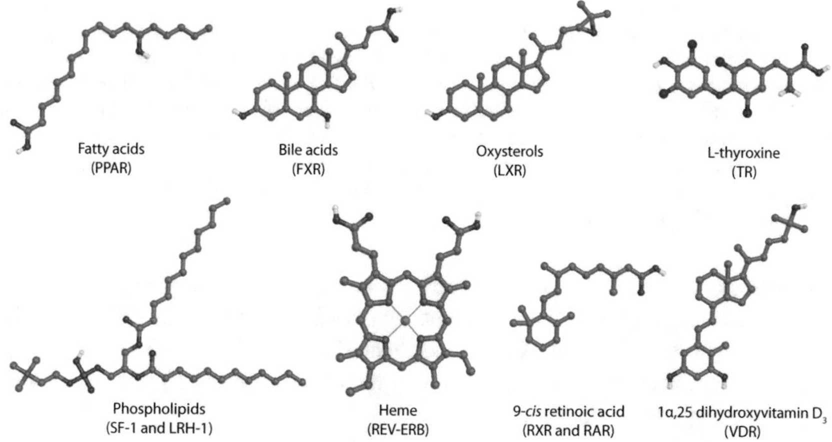

Fig. 3.4 Nuclear receptor ligands. The chemical structure of natural ligands of the nuclear receptors RXR, VDR, FXR, LXR and PPARγ are shown

micro- and macronutrients. In contrast, some additional nuclear receptors, such as hepatocyte nuclear factor (HNF) 4α and 4γ, liver receptor homolog-1 (LRH-1), Reverse-Erb (REV-ERBα, β), RAR-related orphan receptor (RORα, β, γ) and steroidogenic factor 1 (SF-1), bind nutrient derivatives, such as fatty acids, phospolipids, heme and sterols, but this interaction is constitutively and does not represent a sensing process.

All true nutrient sensing nuclear receptors form heterodimers with the sensor for 9-*cis* retinoic acid, retinoid X receptor (RXRα, β, γ), and bind to specific nucleotide sequences, referred to as response elements (REs) (Fig. 3.5), however, some nuclear receptors form homodimers or contact DNA even as monomers. RXR heterodimer complexes permanently locate in the nucleus, i.e. in contrast to GR and AR they do not have first to dissociate from chaperone proteins and then to translocate into the nucleus. This indicates that the macro- and micronutrient sensing by nuclear recep-

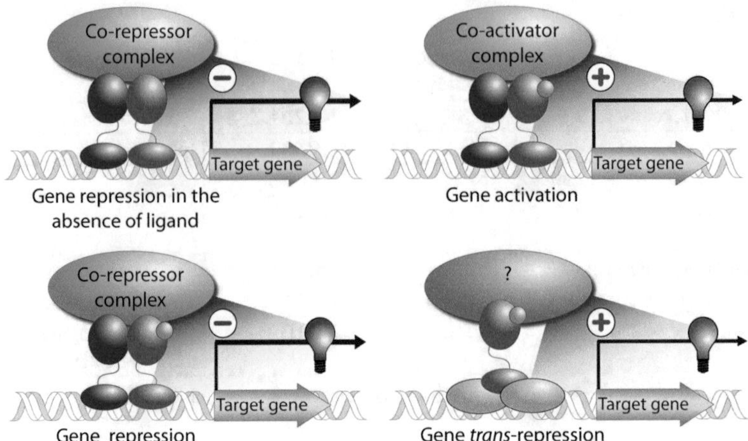

Fig. 3.5 Nuclear receptor action. Genome-wide RXR heterodimers bind specific DNA sequences already in the absence of ligand. In this constellation they mainly interact with co-repressor proteins and are involved in gene repression. After ligand binding nuclear receptors either change their co-factors to co-activators and get involved in gene activation or keep the contact with co-repressors and further repress some of their target genes. As a phenomenon, referred as trans-repression, the nuclear receptor does not bind directly to DNA but interferes with the activity of other transcriptional factors, such as nuclear factor-κB (NF-κB) by protein-protein interactions combined with post-translational modifications

tors takes place in the nucleus. The nutrients act as switches of gene regulation by inducing a conformational change to the ligand-binding domains of their specific nuclear receptors. This results in the coordinated dissociation of co-repressors and the recruitment of co-activator proteins, in order to enable transcriptional activation of up to 1,000 genes (Box 3.1).

Box 3.1 Genome-Wide Nuclear Receptor Analysis

The different next-generation sequence methods, such as ChIP-seq, which were intensively used by the *ENCODE Project* (Box 2.4), have also been applied for the genome-wide analysis of the action of nuclear receptors. The total sum of individual binding sites for an individual nuclear receptor, referred to as its cistrome, collectively for multiple tissues can exceed 20,000. Moreover, transcriptome-wide methods, such as RNA-seq, identified more than 1,000 primary target genes for most nuclear receptors or their specific ligands. Not all of these binding sites and target genes are equally important, but their huge numbers indicate that nuclear receptors and their ligands are involved in the control of more physiological processes than formerly assumed. On many, if not on all of their genomic binding sites nuclear receptors co-locate with other transcription factors, such as spleen focus forming

(continued)

Box 3.1 (continued)
virus proviral integration oncogene (SPI1, also called PU.1), FOXA1 or NF-κB, that either work as pioneer factors to open the local chromatin structure or to interact with other signal transduction pathways, of which these proteins are the end points. In addition, nuclear receptors do not directly contact DNA on all of their genomic binding sites but can sometimes act as co-factors to other DNA-binding transcription factors (Fig. 3.5).

Members of the nuclear receptor superfamily are involved in the regulation of nearly all physiological processes of the human body. Since they represent a class of transcription factors that can easily and very specifically be regulated by small lipophilic compounds, these receptors and their natural ligands play an important role in the maintenance of a steady state of the human body, which is representing "health". The evolutionary oldest and probably still the most important role of nuclear receptors is the control of metabolism. Figure 3.6 illustrates the inter-relationship of lipid metabolism, supplemented by micro- and macronutrients taken up by diet and represented by metabolites and their converting enzymes, such as cytochrome P450s (CYPs), transporters and key representatives of the nuclear

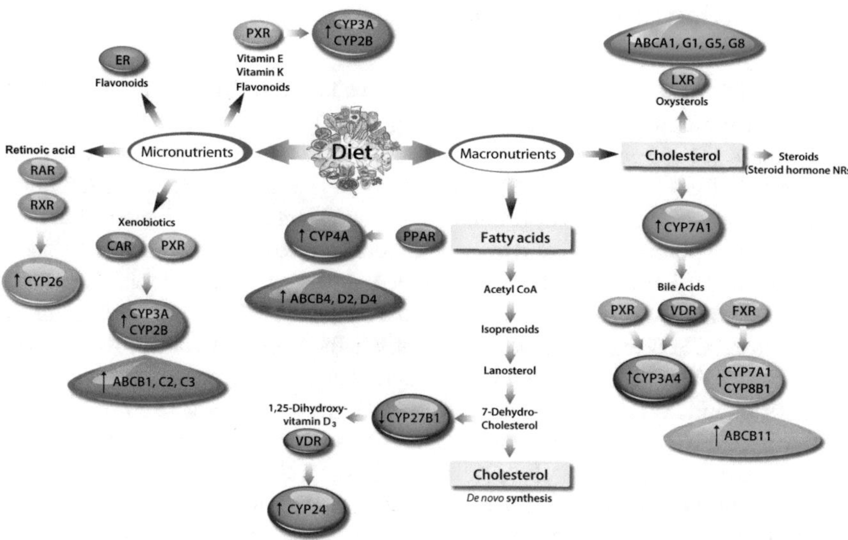

Fig. 3.6 The role of nuclear receptor in lipid metabolism, metabolite enzymes and transporters. The inter-relationship between micro- and macronutrient metabolism involves metabolite enzymes, transporters and nuclear receptors. Only a selected number of metabolites and proteins are shown. Differently color-coded there are many examples of triangle relationships between a metabolite acting as a ligand for a nuclear receptor, nuclear receptors activating their target genes, some of which are metabolic enzymes and transporters for the metabolite. In this feedback loop the metabolite regulates its nuclear receptor, the receptor its enzyme and the enzyme its metabolite

receptor superfamily. There are many examples (RAR, CAR, PXR, PPAR, VDR, LXR and FXR, differently color-coded in Fig. 3.6), where a metabolite activates a nuclear receptor, which in turn controls the expression of the enzyme or transporter handling the metabolite. Nuclear receptor-controlled CYP enzymes have also a central role in receptor ligand inactivation and clearance. These triangle regulatory circuits are found at several critical positions in lipid metabolism pathways and allow a fine-tuned control on metabolite concentrations and nuclear receptor activity. This suggests that dietary metabolites are ancestral precursors of endocrine signaling molecules, such as steroid hormones. In turn this emphasizes the nutrigenomics principle (Chap. 4), that diet is not only a supply for energy, but has also important signaling function.

An immediate implication that followed from understanding the function of nuclear receptors is their potential as therapeutic targets. In fact, nuclear receptor targeting drugs are widely used and commercially successful. For example, bexarotene and alitretinoin (RXRs), fibrates (PPARα), and thiazolidinediones (PPARγ) are already approved drugs for treating cancer, hyperlipidemia and T2D, respectively. Moreover, FXR and LXR agonists are in development for treating non-alcoholic fatty liver disease (NAFLD) (Sect. 9.4) and preventing atherosclerosis (Sect. 11.2).

3.3 Functions and Actions of PPARs

The different steps in handling fatty acids, especially resorbing them in the intestine, transforming them in the liver, burning them in active tissues and collecting their excess for long-term storage in adipose tissue, is coordinated by the three members of the PPAR family of nuclear fatty acid sensors (Fig. 3.7). Common dietary fats, such as oleic acid, LA and ALA, can all act as PPAR ligands, but their respective receptors also respond to various lipid metabolites, such as prostaglandin J2, 8S-hydroxyeicosatetraenoic acid and a number of oxidized phospholipids, respectively. Due to its distinct tissue distribution, each PPAR subtype has unique functions. PPARα is expressed pre-dominantly in the liver, heart and brown adipose tissue (BAT). PPARδ (also called PPARβ in rodents) is ubiquitously expressed but has most important functions in skeletal muscle, liver and heart. In adipose tissue, PPARγ is highly expressed in adipocytes and acts there both as a master regulator of adipogenesis (Sect. 8.2) and a potent modulator of lipid metabolism and insulin sensitivity. Due to alternative splicing and differential promoter usage, there are two PPARγ isoforms, of which PPARγ1 is expressed in many tissues, while the expression of PPARγ2 is restricted to adipose tissue.

After a meal, PPARα and PPARδ are sensing increasing levels of fatty acids efflux from the liver and start to manage lipid metabolism via the promotion of fatty acid β-oxidation and ATP production in mitochondria of skeletal muscles and the heart. PPARα is also the molecular target of fibrates, which are widely used drugs that reduce serum triacylglycerols through the increased oxidation of fatty acids. Also PPARδ can decrease serum triacylglycerols, prevent high-fat diet-induced

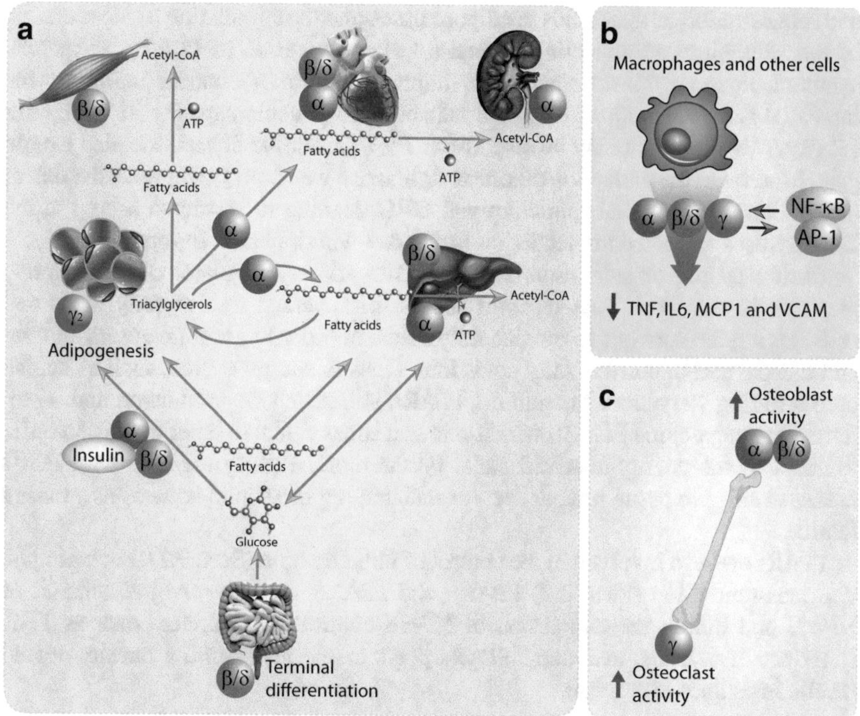

Fig. 3.7 Physiological roles of PPARs. (**a**). PPARα regulates the expression (i) of enzymes that lead to the mobilization of stored fatty acids in adipose tissue and (ii) of fatty acid catabolizing enzymes in the liver, heart and kidney. PPARδ is expressed at high levels in the intestine where it mediates the induction of terminal differentiation of epithelium. Activating PPARδ or PPARγ can increase insulin sensitivity. PPARδ regulates the expression of fatty acid catabolizing enzymes in skeletal muscle where released fatty acids are oxidized to generate ATP. PPARγ promotes the differentiation of adipocytes. (**b**). PPARα, PPARδ and PPARγ can interfere with NF-κB and activating protein 1 (AP-1) in macrophages, endothelial cells, epithelial cells and other tissues, causing the attenuation of pro-inflammatory signaling by decreasing the expression of pro-inflammatory cytokines, chemokines and cell adhesion molecules. (**c**). The activation of PPARα and PPARδ promotes osteoblast activity in bone, whereas the activation of PPARγ promotes osteoclast activity

obesity and increase insulin sensitivity through the regulation of genes encoding fatty acid metabolizing enzymes in skeletal muscle and genes for lipogenic proteins in the liver. Moreover, PPARδ increases serum HDL-cholesterol levels via stimulating the expression of the reverse cholesterol transporter ATP-binding cassette (ABC) A1 and apolipoprotein (APO) A1-specific cholesterol efflux. PPARγ promotes adipose tissue differentiation together with fibroblast growth factors (FGFs) 1 and 21, in order to store excess of non-consumed fatty acids.

In addition, in adipose tissue PPARγ also controls glucose uptake via the regulation of *SLC2A4* expression (encoding for GLUT4). High concentrations of circulating fatty acids can cause insulin resistance (Sect. 9.4). Therefore, enhanced uptake and sequestration of fatty acids in adipose tissue and the stimulation of the secretion of the adipokines adiponectin and resistin after activation of PPARγ by specific synthetic ligands, thiazolidinediones, improve insulin resistance. However, the increased fatty acids uptake and the enhanced adipogenic capacity in WAT after PPARγ activation both may be responsible for thiazolidinedione-associated weight gain. Moreover, the thiazolidinedione rosiglitazone was found to increase the risk of heart failure, myocardial infarction and CVD, leading to restricted access in the United States and a recommendation for market withdrawal in Europe.

During fasting or starvation, PPARα is the primary regulator of the adaptive response in the liver. This receptor senses the reversed flux of fatty acids and activates a gene network to oxidize fatty acids, in order to generate energy in liver and muscle and to convert fatty acids into a usable energy source, such as ketone bodies during starvation. In addition, PPARα stimulates the production and secretion of the hepatokine FGF21, which acts as a stress signal to other tissues, in order to adapt to the energy-deprived state. Furthermore, PPARγ together with FGF1 mediates adipose tissue remodeling for maintaining metabolic homeostasis during famine.

PPARs are also involved in the control of inflammation (Sect. 7.2). For example, via trans-repression (Sect. 3.2) PPARα and PPARδ sequester the p65 subunit of NF-κB and inhibit the expression of NF-κB-controlled cytokines, such as TNF, IL1B and IL6. Thus, activating PPARs plays a role in inhibiting obesity-related insulin resistance (Sect. 9.4).

3.4 Integration of Lipid Metabolism by LXRs and FXR

LXRs and FXR are sensors for the cholesterol derivates oxysterols and bile acids, respectively. These nuclear receptors do not only regulate cholesterol and bile acid metabolism but also have a central role in the integration of sterol, fatty acid and glucose metabolism. LXRα is expressed in tissues with a high metabolic activity, such as liver, adipose and macrophages, whereas LXRβ is found ubiquitously. LXRs, similarly to PPARs, have a large hydrophobic ligand-binding pocket that can bind to a variety of different ligands, such as 24(S)-hydroxycholesterol, 25-hydroxycholesterol, 22(R)-hydroxycholesterol and 24(S),25-epoxycholesterol, at their physiological concentrations. FXR is expressed mainly in the liver, intestine, kidney and adrenal glands. Bile acids, such as chenodeoxycholic acid and cholic acid, are endogenous FXR ligands, but they can also activate the nuclear receptors PXR, CAR and VDR.

LXR may be known best for its ability to promote reverse cholesterol transport (Sect. 7.3), i.e. cholesterol delivery from the periphery to the liver for excretion (Figs. 3.8 and 7.5). This involves the transfer of cholesterol to APOA1 and pre-β HDLs via the transporter protein ABCA1, which is encoded by one of the most prominent LXR target genes. Further important LXR targets are ABCG1, which together with ABCA1 promotes cholesterol efflux from macrophages, and the intra-

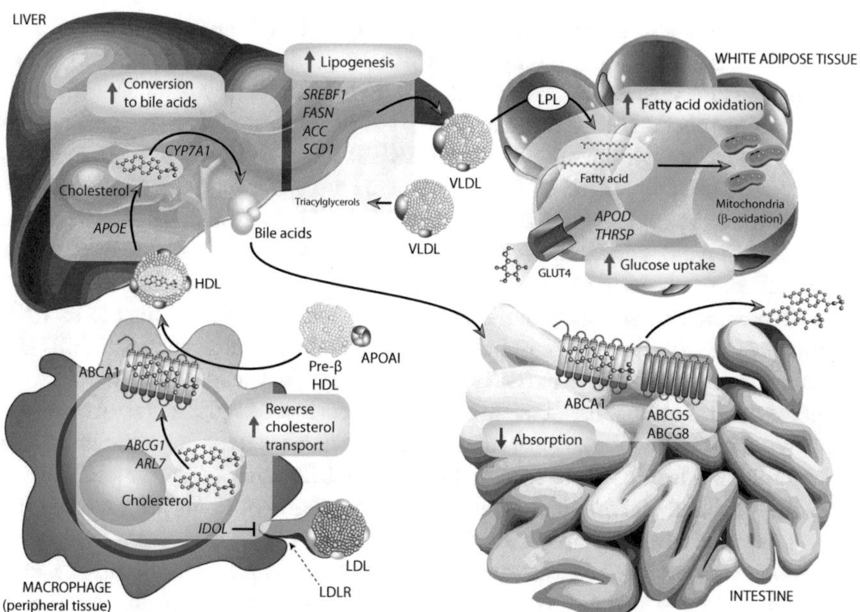

Fig. 3.8 Effects of LXR on metabolism. LXR has effects on multiple metabolic pathways. In macrophages, LXR induces the expression of the genes *IDOL*, *ARL7*, *ABCA1* and *ABCG1*. In the liver, LXR promotes fatty acid synthesis via induction of *SREBF1* and its targets, *FASN*, *ACC* and *SCD1*. Triglyceride-rich very low-density lipoproteins (VLDLs) in the liver serve as transporters for lipids to peripheral tissues, including adipose tissue, where the action of LPL liberates fatty acids from VLDLs. In adipose tissue, LXR regulates the expression of *APOD* and *THRSP* and promotes the breakdown of fatty acids via β-oxidation and glucose uptake via induction of GLUT4. Finally, in the intestine, LXR inhibits cholesterol absorption by inducing the expression of the ABCG5-ABCG8 complex

cellular trafficking protein ADP-ribosylation factor-like 4C (ARL4C) that facilitates cholesterol delivery to the plasma membrane. LXR also down-regulates the gene of the LDL receptor (*LDLR*) and up-regulates the gene inducible degrader of LDLR (*IDOL*). Thus, activation of LXR attenuates LDL uptake by macrophages.

In the liver, LXR induces the expression of a cluster of apolipoprotein genes including *APOE*, *APOC1*, *APOC2* and *APOC4*, being involved in lipid transport and catabolism. LXR also up-regulates genes of lipid remodeling proteins, such as phospholipid transfer protein (*PLTP*), cholesterol ester transfer protein (*CETP*) and lipoprotein lipase (*LPL*) (Fig. 3.8). Furthermore, LXR induces the expression of several genes that mediate the elongation and the unsaturation of fatty acids, which leads to synthesis of long-chain PUFAs, such as ω-3 fatty acids. This increase in long-chain PUFAs results in decreased expression of NF-κB target genes via epigenetic silencing of their regulatory regions. Long-chain PUFAs also function as substrates for the enzymes that synthesize eicosanoids and specialized inflammation-resolving lipid mediators, such as resolvins and protectins.

LXR also induces the gene for the enzyme lysophospholipid acyltransferase 3 (LPCAT3), thus mediating the synthesis of phospholipids containing long-chain PUFAs. This decreases ER stress (Sect. 7.5) and inflammatory responses. In addition, a major function of LXR in the liver is the promotion of *de novo* biosynthesis of fatty acids through the stimulation of SREBF1, ACC, FASN and steroyl-CoA desaturase 1 (SCD1). Some of these fatty acids are esterified with cholesterol, in order to avoid toxic levels of free cholesterol. In the intestine, LXR induces the expression of genes encoding the transporters ABCG5 and ABCG8 that mediate the apical efflux of cholesterol from enterocytes. In adipose tissue, LXR also affects glucose metabolism via the stimulation of GLUT4 expression. In this tissue, LXR regulates the expression of lipid-binding and metabolic proteins, such as APOD and thyroid hormone responsive (THRSP), and increases fatty acid β-oxidation.

In the control of lipid metabolism FXR often acts in a complementary or reciprocal way to LXR (Fig. 3.9). Since high bile acid concentrations are toxic to cells, a central function for FXR is to control these levels. Bile acids are cholesterol derivatives that facilitate the efficient digestion and absorption of lipids after a meal, but they represent also the major way to eliminate cholesterol from the body. Nevertheless, most bile acids are recycled via the enterohepatic circulation, i.e. they pass from the intestine back to the liver. FXR controls this bile acid flux via modulating their synthesis, modification, absorption and uptake. In the liver, FXR inhibits bile acid synthesis by repressing the enzymes CYP7A1 and CYP8B1. FXR stimulates the secretion of FGF19 from the intestine, which then activates FGF receptor 4 (FGFR4) in the liver and in this way provides a complementary mechanism for the feedback inhibition of bile acid synthesis.

In the gall bladder, FXR up-regulates the enzymes bile acid-CoA synthase (SLC27A7) and bile acid-CoA-amino acid N-acetyltransferase (BAAT), which catalyze the conjugation of bile acid with the amino acids taurine or glycine, and the bile salt export pumps ABCB11 and ABCB4. In the intestine, FXR inhibits the absorption of bile salts via the down-regulation of the apical sodium-dependent bile salt transporter SLC10A2 and promotes the movement of bile salts from the apical to the basolateral membrane of enterocytes via the up-regulation of ileal fatty acid-binding protein (FABP6). FXR limits hepatic bile salt levels by the down-regulation of the bile acid transporters solute carrier organic anion transporters (SLCOs) A1 and A2. Furthermore, in order to protect the liver from toxicity, FXR induces the expression of the enzymes CYP3A4 and CYP3A11, which hydroxylate bile acids, as well as sulfotransferase family 2A, member 1 (SULT2A1) and UDP glucuronosyltransferase 2 family, polypeptide B4 (UGT2B4), which sulphate and glucuronidate bile acids, respectively. In addition, in the liver FXR reduces lipogenesis via the repression of *SREBF1* and *FASN* gene expression, i.e the receptor decreases triacylglycerole levels. Finally, FXR also influences hepatic carbohydrate metabolism and inhibits glucogenesis via the down-regulation of glucose-6-phosphatase (G6PC) and phosphoenolpyruvate carboxykinase (PCK) 2.

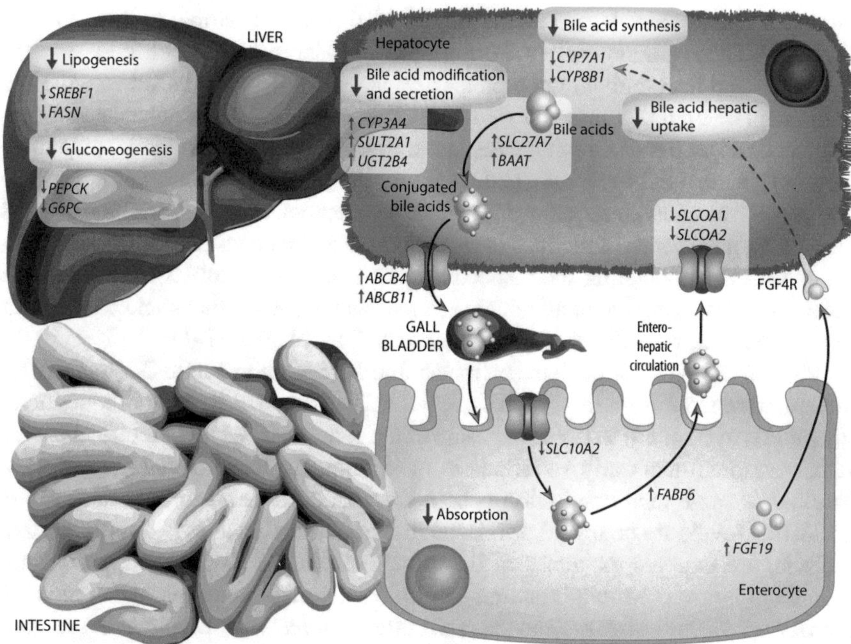

Fig. 3.9 Effects of FXR on metabolism. In the liver, FXR reduces the conversion of cholesterol to bile acids by down-regulating the enzymes CYP7A1 and CYP8B1. FXR also reduces bile acid toxicity in the liver by increasing *SULT2A1*, *UGT2B4* and *CYP3A4* expression. Bile acids are conjugated to either glycine or taurine before secretion into the bile, which is enhanced by FXR via increasing *SLC27A7* and *BAAT* expression. Moreover, FXR promotes the transport of bile acids to the gall bladder via up-regulation of the bile salt export pumps ABCB11 and ABCB4. Within the intestine, FXR reduces bile acid absorption via down-regulation of the apical sodium-dependent bile acid transporter SLC10A2 and promotes bile acid movement across the enterocyte via controlling FABP6. In the liver, FXR reduces hepatic uptake of bile acids by reducing the expression of the transporters SLCOA1 and SLCOA2. FXR also promotes the release of FGF19 from the intestine, which acts on FGF4R to reduce CYP7A1 expression and thus represses bile acid synthesis. In the liver, FXR also acts on glucose metabolism by reducing gluconeogenesis via PCK2 and G6PC and lipogenesis via inhibition of SREBF1 and FASN

3.5 Coordination of the Immune Response by VDR

The immune system is composed by a multitude of highly specialized cells (Box 2.2) that are all created by a differentiation process, referred to as hematopoiesis. Cellular differentiation is controlled by epigenetic mechanisms (Chap. 5), in which a number of transcription factors play a key role. Cells of the immune system have a rapid turnover and are therefore able to show a maximal adaptive response to environmental changes. Lipid sensing and signaling via nuclear receptors has an important role in the differentiation and subtype specification of immune cells, such as T cells, macrophages and dendritic cells (Chap. 7). Importantly, these cells are very mobile and are found in a wide range of subtypes nearly everywhere in the human body, i.e. also in metabolic tissues in disease scenarios, such as obesity (Chap. 8). These immune cells are often exposed to large amounts of lipids, such as pathogen- and host-derived lipoproteins or lipids of apoptotic cells, and are the key integrators of lipid and immune signaling. This means that macrophages and dendritic cells as well as their precursors, monocytes, coordinate metabolic, inflammatory and general stress-response pathways via changes of their transcriptome profile and respective subtype specification. Nuclear receptors, such as VDR, RAR, LXR and PPAR, have central functions in sensing these endogenous and exogenous stimuli as well as in adapting the respective gene expression profiles of the immune cells. As a representative of all micro- and macronutrient-sensing nuclear receptors, in the following we will focus on VDR and its ligand $1,25(OH)_2D_3$.

Vitamin D_3 and its most abundant metabolite, 25-hydroxyvitamin D_3 ($25(OH)D_3$), either derive from from diet, such as fatty fish, or from endogenous production of vitamin D_3 in response to UV-B exposure of the skin (Sect. 4.1). Therefore, there are rather seasonal than daily variations in the vitamin D status of the human body. Worldwide more than one billion people are vitamin D deficient, i.e. their serum $25(OH)D_3$ levels are below 50 nM. Bone malformations, such as rickets and osteomalacia, are extreme examples of the effects of vitamin D deficiency, but since vitamin D is involved in a broad range of physiological processes, it can increase the risk for various diseases and susceptibility for infections. Living at higher latitudes, i.e. at significant seasonal variations of UV-B exposure, increases the risk of the autoimmune disease T1D, multiple sclerosis and Crohn's disease, but also of metabolic diseases, such as hypertension, T2D and CVD.

VDR is expressed in all important cell types of the immune system, i.e. these cells are sensitive to changes in the vitamin D concentrations. Importantly, macrophages and dendritic cells express the enzyme CYP27B1 that converts $25(OH)D_3$ into the VDR ligand $1,25(OH)_2D_3$, i.e. in these cells vitamin D can act autocrine or paracrine (Fig. 3.10). Interestingly, while *CYP27B1* expression in the kidneys is negatively regulated by a number of signals, such as Ca^{2+}, parathyroid hormone, phosphate and $1,25(OH)_2D_3$, antigen-presenting cells do not respond to these inhibitory signals but rather further up-regulate *CYP27B1* expression after stimulation with cytokines and TLR ligands.

TLRs and other PRRs detect pathogens on the surface of macrophages and initiate an immune response (Sect. 7.2), for example against the intra-cellular bacterium *Mycobacterium tuberculosis*. Notably, no other infectious disease had so far more human victims (in total up to one billion) than tuberculosis. In macrophages, the TLR-triggered increased expression of VDR target genes encoding for anti-microbial peptides, such as cathelicidin (*CAMP*) and defensin, beta 4A (*DEFB4*), efficiently kills intra-cellular *M. tubercolosis* (Fig. 3.10). This vitamin D-dependent anti-microbial mechanism can explain, why (i) sun or artificial UV-B exposure is efficient in the supportive treatment of tuberculosis, (ii) vitamin D deficiency is associated with tuberculosis, (iii) some variations of the *VDR* gene increase the susceptibility to *M. tuberculosis* infection and (iv) humans with dark skin living distant from the equator have an increased susceptibility to tuberculosis infection. Moreover, vitamin D-induced cytokine production of T cells and monocytes modu-

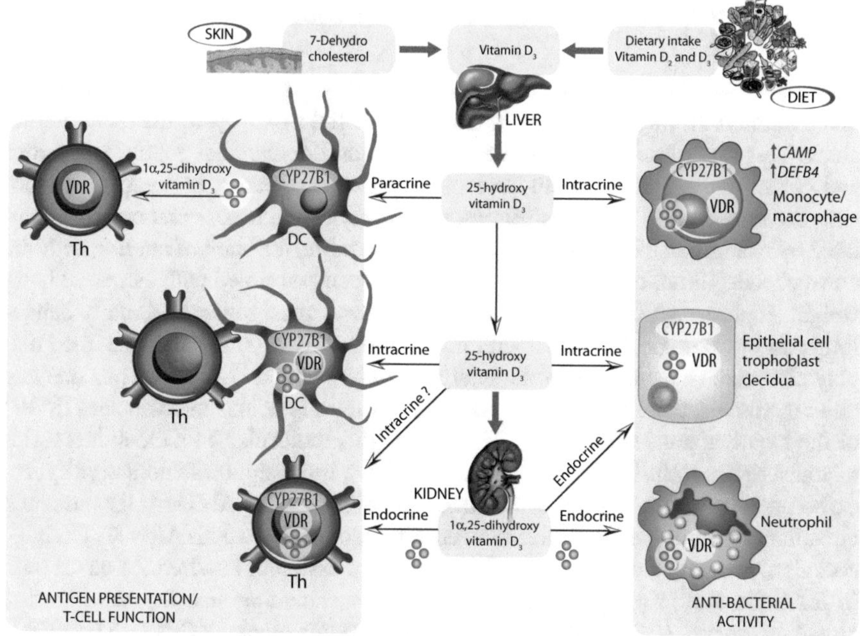

Fig. 3.10 Innate and adaptive immune responses to vitamin D. Macrophages and dendritic cells express the vitamin D-activating enzyme CYP27B1 and VDR can then utilize 25(OH)D_3 for intracrine and paracrine responses via localized conversion to active 1,25$(OH)_2D_3$. In monocytes and macrophages, this promotes the anti-bacterial response to infection via CAMP and DEFB4. 1,25$(OH)_2D_3$ inhibits dendritic cell maturation and modulates T helper (T_H) cell function. Intracrine immune effects of 25(OH)D_3 may also occur in *CYP27B1*/*VDR*-expressing epithelial cells. In contrast, most other cells, such as T_H cells and neutrophils, depend on the circulating levels of 1,25$(OH)_2D_3$ that are synthesized by the kidneys, i.e. they are endocrine targets of 1,25$(OH)_2D_3$

late *CAMP* expression. Taken together, the availability of the micronutrient vitamin D is essential for an appropriate response to infections.

Like many other nuclear receptors, VDR can antagonize, via trans-repression of transcription factors, such as NF-AT, AP-1 and NF-κB, the inflammatory response of immune cells. The resulting decreased expression of cytokines, such as IL2 and IL12, demonstrates the anti-inflammatory potential of vitamin D metabolites. However, VDR is a key transcription factor in the differentiation of myeloid progenitors into monocytes and macrophages (Sect. 7.1). In contrast, in dendritic cells vitamin D inhibits differentiation, maturation and immunostimulatory capacity via the repression of the genes encoding for the different variants of major histocompatibility complex (HLA) and its costimulatory molecules CD40, CD80, CD86 and the up-regulation of inhibitory molecules, such as chemokine (C-C motif) ligand 22 (CCL22) and IL10. This tolerogenic (i.e. immune tolerance inducing) phenotype of dendritic cells is associated with the induction of T_{REG} cells (Fig. 3.10).

3.6 Circadian Control of Metabolic Processes

Artificial light, shift work, travel and temporal disorganization have disrupted for many humans the alignment between the external light-dark cycle and their internal clock. This is of disadvantage for metabolic health. Longitudinal population studies and clinical investigations both have indicated an association between shift work and diseases, such as T2D, gastrointestinal disorders and cancer that can be modulated by changes in the circadian rhythm. Furthermore, the habit of altering bedtime on weekends, the so-called "social jet lag", has been associated with increased body weight. Light-sensitive organisms, such as humans, synchronize their daily behavioral and physiological rhythms with the rotation of Earth on its axis, i.e. they display circadian (meaning "approximately one day") activity cycles. In humans and other mammals, these rhythms are generated by the suprachiasmatic nucleus (SCN) of the hypothalamus (Fig. 3.11a). The core of this molecular 24 h clock is a series of transcription-translation feedback loops of the transcription factors aryl hydrocarbon receptor nuclear translocator-like (ARNTL, also called BMAL1) and clock circadian regulator (CLOCK) and their co-repressor proteins. This ARNTL-CLOCK complex, in a circadian fashion, activates the expression of hundreds of genes both in the brain and in peripheral metabolic tissues, including also the genes period circadian clock 1 (*PER1*) and cryptochrome circadian clock 1 (*CRY1*). The PER1-CRY1 co-repressor complex inactivates ARNTL-CLOCK, but phosphorylation and ubiquitylation of CRY1 during the night initiates the proteosomal degradation of the

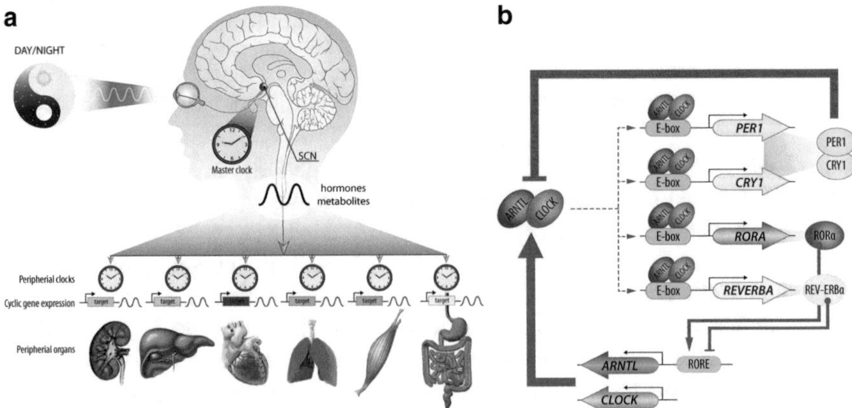

Fig. 3.11 The circadian clock in mammals. (**a**). Electrical and humoral signals from the SCN synchronize phases of circadian clocks in peripheral organs, which then generate time dependent rhythms in gene expression, metabolism and other physiological activities. (**b**). In the feedback loop of the molecular circadian oscillator positive elements, such as the transcription factors ARNTL, CLOCK and ROR, are shown in *red*, and negative elements, such as PER1, CRY1 and REV-ERBα, in *yellow*. The combined actions of hundreds of ARNTL-CLOCK target genes provide a circadian output in physiology

repressors and re-activates ARNTL-CLOCK. The genes encoding for the nuclear receptors REV-ERBα and ROR are further targets of ARNTL-CLOCK. REV-ERBα negatively and ROR positively regulates the expression of the *ARNTL* gene, i.e. these nuclear receptors form additional feedback loops in the control of the molecular clock (Fig. 3.11b).

The molecular clock is a self-sustained activity, but circadian gene transcription can also be modulated by metabolites, in particular by those representing energetic flux (Box 3.2). For example, the AMP sensor AMPK (Sect. 6.5) connects the internal clock function to the nutrient state via phosphorylation and subsequent proteasomal degradation of the ARNTL-CLOCK repressor CRY1 (Fig. 3.12a). In parallel, the cyclical activity of ARNTL-CLOCK is modulated by the HDM lysine-specific demethylase 5A (KDM5A, also called JARID1A), which in turn is linked via its co-factors iron and α-ketoglutarate to cellular redox and mitochondrial energetics. The bidirectional interaction between circadian and metabolic signaling is the inhibition of ARNTL-CLOCK by the nicotinamide adenine dinucleotide (NAD^+)-dependent histone deacetylase (HDAC) SIRT1 (Fig. 3.12b). This represents another feedback control of the molecular clock, since the gene encoding for the critical enzyme for NAD^+ synthesis, nicotinamide mononucleotide phosphoribosyltransferase (*NAMPT*, also known as visfatin), is a direct ARNTL-CLOCK target.

Since NAD^+-dependent sirtuins are important regulators of metabolic pathways in response to calorie restriction (Sect. 6.4), the link between the molecular clock

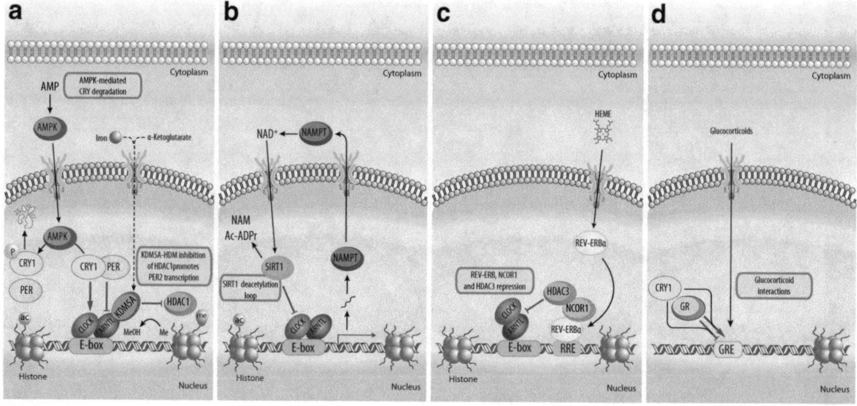

Fig. 3.12 Links between circadian and metabolic regulation. (**a**). The ARNTL-CLOCK complex is supported by the α-ketoglutarate-triggered HDM KDM5A, while AMP-stimulated AMPK phosphorylates the repressor protein CRY1 and controls its proteolytic degradation. (**b**). The ARNTL-CLOCK heterodimer up-regulates the *NAMPT* gene that encodes for an enzyme critical for NAD^+ synthesis. This increases NAD^+ and the activity of SIRT1, which itself is a negative feedback regulator of the ARNTL-CLOCK complex. (**c**). The repressive complex of REV-ERBα with HDAC3 and NCOR1 is sensitive to concentrations of heme levels and inactivates the ARNTL-CLOCK complex. (**d**). Binding of GR for its genomic loci is triggered by glucocorticoids and CRY1

Box 3.2 Modulating the Molecular Clock Metabolic by Systems

The proteins of the molecular clock are not only expressed in the SCN but also at peripheral sites, such as the liver. Entrainment of the liver clock to feeding involves glucocorticoid signaling, temperature via heat shock transcription factor 1 (HSF1) and ADP-ribosylation. In this way, the circadian clock coordinates daily behavioral cycles of sleep-wake and fasting-feeding with anabolic and catabolic processes in the periphery. The central and peripheral clocks are also synchronized via post-translational modifications of transcription factors and histones that tune gene expression rhythms into changes of the metabolic state. Therefore, in addition to the transcription-translation-based feedback systems, mammals and other species use NAD^+ oscillation, redox flux, ATP availability and mitochondrial function, in order to influence acetylation and methylation reactions (Fig. 3.13). For example, the redox-based clock represents oscillations in the redox state of the family of peroxiredoxin anti-oxidant enzymes that rhythmically anticipate the generation of ROS. Moreover, NAD^+ is an electron shuttle in oxidoreductase reactions and also acts as a co-factor in HDAC and ADP-ribosylation modifications (Sect. 5.2).

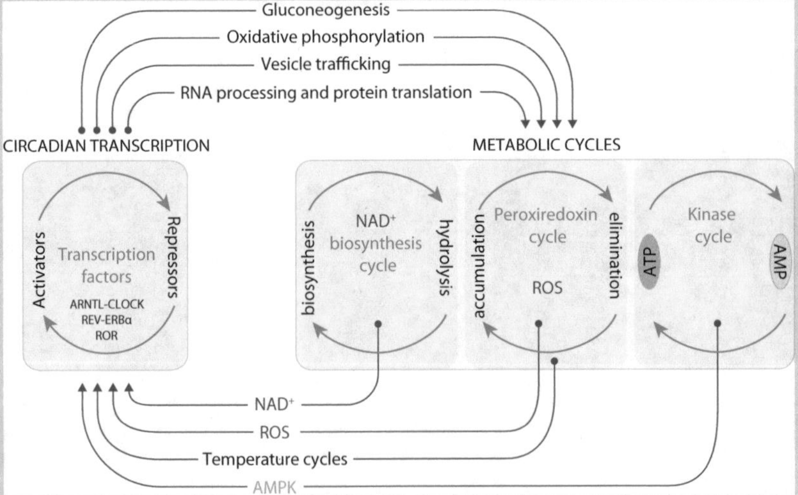

Fig. 3.13 Cross-talk between circadian transcription and metabolic systems. The key transcription factors ARNTL-CLOCK, REV-ERBα and ROR control cellular and metabolic pathways, such as gluconeogenesis, oxidative phosphorylation, vesicle trafficking and RNA processing and translation, in a cyclical fashion. Metabolic cycles reciprocally affect this molecular clock. The cycles of NAD^+ biosynthesis as well as the activity of peroxiredoxin and various kinases (for example, AMPK, see Fig. 3.12) generate active intermediates that provide feedback to regulate the core clock transcriptional network

and sirtuin activity has implications for aging. A further example is the circadian recruitment of HDAC3 to the complex of REV-ERBα with the co-repressor NCOR1, which leads to the rhythmic repression of genes involved in lipogenesis and carbohydrate metabolism (Fig. 3.12c). Finally, many nuclear receptors exhibit, at least in mice, circadian oscillation in their expression, so that the timing of the interaction with their specific ligands allows coupling between temporal and physiological systems (Fig. 3.12d). Interestingly, the expression of REV-ERBα is also modulated by glucocorticoids.

Future View

The future, further insight on nutrient sensing systems, including that via nuclear receptors, will allow a more integrative view of the molecular (re)action of the human body to dietary molecules. This will not only address the cross-regulation between different nutrient-sensing pathways, but will also incorporate other signaling pathways, such as those ones controlling cellular growth or mediating chronic inflammation. The dense integration of metabolic pathways makes it difficult to achieve specific responses to a treatment with one natural or synthetic nuclear receptor ligand without provoking side effects through compensatory or complementary responses from other pathways. Nevertheless, future studies that will be performed in a safe and clever way in humans (rather than in rodents) should provide a level of understanding that will lead to more specific ligands and identification of new target genes regulated by dietary components. Moreover, a better understanding of the links between circadian biology and metabolism will allow tailoring preventive interventions and therapies. Taken together, the field of personalized nutrition/medicine will benefit a lot from this additional insight.

Key Concepts

- Periodical scarcity of nutrients was a strong evolutionary pressure to select efficient mechanisms of their sensing.
- Fatty acids are sensed by GPRs in the plasma membrane and cholesterol by the SCAP-SREBF complex in the ER membrane.
- LXR is activated by elevated cholesterol levels, i.e. the pathways of SREBF and LXR work in a reciprocal fashion, in order to maintain cellular and systemic cholesterol homeostasis.
- The TOR component mTORC1 senses amino acids, which are scavenged in lysosomes from cellular components through autophagy.
- The transporter GLUT2 and the hexokinase GCK are true sensors for glucose, since they are only active at high but not at low physiologic glucose concentrations.
- Many nuclear receptors bind micro- and macronutrients or their metabolites with an affinity that varies between 0.1 nM for VDR and up to mM for PPARs and reflect the physiological concentrations of the respective molecules.

- Micro- and macronutrients act as switches to gene regulation by inducing a conformational change to the ligand-binding domain of their specific nuclear receptors.
- The evolutionary oldest and probably still the most important role of nuclear receptors is the control of metabolism. This also implies that dietary metabolites are ancestral precursors of endocrine signaling molecules.
- The different steps in handling fatty acids, such acquiring them in the intestine, transforming and delivering them from the liver, burning them in active tissues and collecting their excess for long-term storage in adipose tissue, is coordinated by the three members of the PPAR family.
- PPARγ is most highly expressed in adipose tissue, where it is a master regulator of adipogenesis as well as a potent modulator of whole-body lipid metabolism and insulin sensitivity.
- During fasting or starvation PPARα is the primary regulator of the adaptive response in the liver, where the receptor senses the reversed flux of fatty acids and activates a gene network to convert fatty acids into a usable energy source, such as ketone bodies.
- PPARα and PPARδ sequester through trans-repression the p65 subunit of NF-κB. The reduced activity of NF-κB-controlled leads to reduced expression of cytokines, such as TNF, IL1B and IL6, and less hepatic inflammation caused by dietary and chemical stimuli.
- LXRs and FXR are sensors for the cholesterol derivates oxysterols and bile acids, respectively.
- LXR promotes reverse cholesterol transport, i.e. cholesterol delivery from the periphery to the liver for excretion.
- FXR controls the bile acid flux via modulating their synthesis, modification, absorption and uptake.
- The nuclear receptors VDR, RAR, LXR and PPAR sense endogenous and exogenous stimuli of immune cells and adapt the gene expression profiles of these cells.
- In macrophages the TLR-triggered increased expression of VDR target genes encoding for anti-microbial peptides, such as *CAMP* and *DEFB4*, leads to the efficient killing of intra-cellular *M. tuberculosis* bacteria.
- VDR is a key transcription factor in the differentiation of myeloid progenitors into monocytes and macrophages.
- The core of the molecular clock is a series of transcription-translation feedback loops of the transcription factors ARNTL and CLOCK and their co-repressor proteins PER1 and CRY1.
- The nuclear receptors REV-ERBα and ROR are targets of ARNTL-CLOCK. REV-ERBα negatively and ROR positively regulates the expression of the *ARNTL* gene, i.e. they form additional feedback loops in the control of the molecular clock.
- AMPK couples the molecular clock to the nutrient state via phosphorylation and subsequent proteasomal degradation of the ARNTL-CLOCK repressor CRY1.
- The link between the molecular clock and sirtuin activity may have implications for aging.

Additional Reading

Ahmadian M, Suh JM, Hah N, Liddle C, Atkins AR, Downes M, Evans RM (2013) PPARγ signaling and metabolism: the good, the bad and the future. Nat Med 19:557–566

Bass J (2012) Circadian topology of metabolism. Nature 491:348–356

Calkin AC, Tontonoz P (2012) Transcriptional integration of metabolism by the nuclear sterol-activated receptors LXR and FXR. Nat Rev Mol Cell Biol 13:213–224

Carlberg C, Molnár F (2014) Mechanisms of gene regulation. Springer, Dordrecht. ISBN 978-94-007-7904-4

Efeyan A, Comb WC, Sabatini DM (2015) Nutrient-sensing mechanisms and pathways. Nature 517:302–310

Evans R, Mangelsdorf D (2014) Nuclear receptors, RXR, and the big bang. Cell 157:255–266

Nagy L, Szanto A, Szatmari I, Szeles L (2012) Nuclear hormone receptors enable macrophages and dendritic cells to sense their lipid environment and shape their immune response. Physiol Rev 92:739–789

Chapter 4
Adaption of the Human Genome to Dietary Changes

Abstract Nutrition is essential for life, but the effects of nutritional molecules are complex and influenced by many factors. Genes influence the dietary response, while nutrients, or the lack of them, can affect gene expression. More than 90 % of human genes have not changed since the life in the stone ages, where food availability meant survival. Humans evolved a sense for taste, in order to detect the most energy-rich diet, but this initial survival instinct nowadays causes overweight and obesity. Modern nutrition research has taken up many elements from molecular biology and next-generation sequencing technologies and turned into nutrigenomics. This new discipline attempts to understand the effects of food on multiple molecular levels, such as genomics and epigenomics.

In this chapter, we will start with a definition of nutrigenomics and will provide a few examples of its applications and potential. We will discuss the molecular basis for the recent adaption of the human genome to environmental changes, such as less UV-B exposure after migrating north, and dietary challenges due to dairy farming, such as lactose tolerance. We will demonstrate that the majority of disease-associated genetic variants are located outside of protein-coding regions, for example, within transcription factor binding sites. This means that regulatory SNPs are rather the rule than the exception. Nutrigenomics will be understood as the application of various "omics" technologies for investigations on the level of the epigenome, genome, transcriptome, proteome and metabolome. We will experience that nowadays it is possible to apply these methods for a most comprehensive assessment of a human individual, which is summarized as integrative personal omics profile (iPOP). These individual datasets will be the basis for the optimization of personalized nutrition for preserving health via the prevention of disease, such as T2D and CVD.

Keywords Nutrigenomics • Gene expression • Skin color • Vitamin D • Amylase • Positive selection • Lactose tolerance • Regulatory SNPs • Omics technologies • Integrative personal omics profile

C. Carlberg et al., *Nutrigenomics*, DOI 10.1007/978-3-319-30415-1_4

4.1 Definition of Nutrigenomics

In nutrigenomics, high-throughput "omics" technology is applied in nutrition research (Sect. 4.5). Human diet is a complex mixture of biologically active molecules that can be both micro- and macronutrients (Sect. 1.1). Some of these molecules can (i) have a direct effect on gene expression (A in Fig. 4.1), (ii) modulate, after being metabolized, the activity of a transcription factor (B in Fig. 4.1) or (iii) stimulate a signal transduction cascade that ends with the induction of a transcription factor (C in Fig. 4.1). Nutrigenomics aims to describe, characterize and integrate these interactions between diet and gene expression genome-wide. The results of these investigations lead to an improved understanding of how nutrition influences metabolic pathways and homeostatic control. This nutrition-triggered regulation may be disturbed in the early phase of a diet-related disease, such as T2D. When human individuals are classified according to the interplay of their lifestyle, metabolic pathways and genetic variation, the molecular insight based on nutrigenomics studies can suggest tailored diets, referred to as personalized nutrition, for early therapeutic intervention. For example, human individuals that are genetically at risk for T2D, but not yet in a pre-diabetic state, may be recommended a customized diet, in order to avoid developing the disease. Using diet for a specific therapy dissolves the distinction between food and drugs as well as the definition of health and disease. Therefore, the best advice for healthy eating may result in a more individualized lifestyle.

Nutrigenomics and pharmacogenomics have some similarities concerning concepts and methodological approaches. However, in pharmacogenomics the effects of a single clearly defined compound (a drug) of a precise concentration and a specific target are investigated, whereas nutrigenomics faces the complexity and variability of nutrition. However, some nutritional compounds can reach up to mM concentrations without becoming toxic, while drugs act at clearly lower concentrations.

4.2 Vitamin D and Skin Color

Vitamin D is a micronutrient that can be obtained in its D_3 form from animals, such as fatty fish, or as D_2 from plants or fungi (Fig. 4.2). However, at present as well as in the stone ages, the average human diet is not providing sufficient amounts of the vitamin. Therefore, for humans it is critical to use their own vitamin D_3 production pathway, which starts in the skin from 7-dehydrocholesterol and uses essentially the energy of UV-B radiation from sunlight (Fig. 4.2). The amount of vitamin D_3 synthesis in the skin depends on the UV-B dose received that varies based on season, day length, latitude, altitude and out-door activities of the individual. Moreover, endogenous vitamin D_3 production depends on the melanin content of the skin, i.e. darker skin color can be a very effective natural sunscreen. Two hydroxylation steps

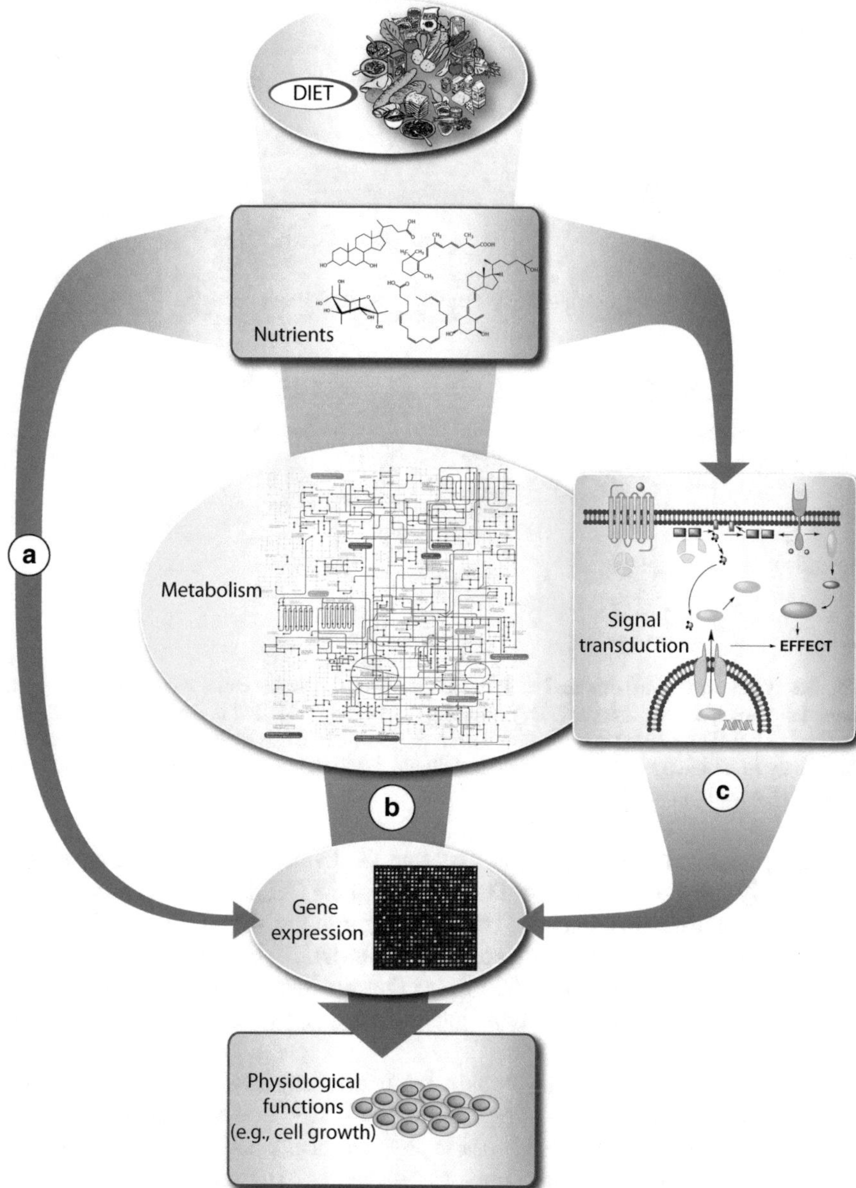

Fig. 4.1 Basis of nutrigenomics. Nutrigenomics seeks to provide a molecular understanding for how dietary nutrients affect health by altering the expression of a larger set of genes. These nutritional compounds have been shown to alter gene expression in a number of ways. For example, they may (**a**) act as direct ligands for transcription factors, (**b**) be transcription factor modulators after a chemical conversion in primary or secondary metabolic pathways or (**c**) serve as activators of signal transduction cascades that end with the activation of a transcription factor. All three activation pathways modulate physiological effects, such as cellular growth

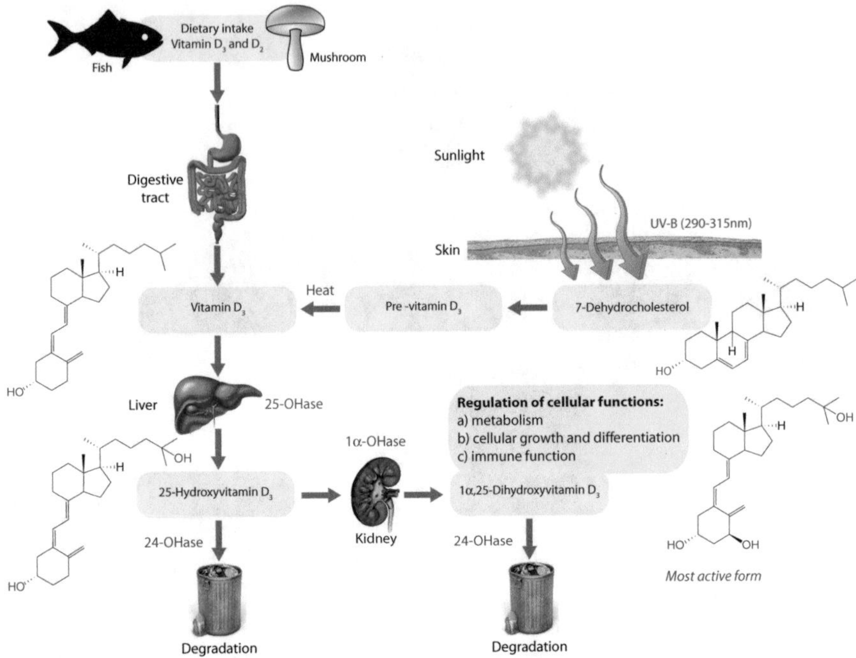

Fig. 4.2 Vitamin D synthesis and metabolism pathway. Humans can obtain vitamin D_3 from animal diet, such as fatty fish (or in D_2 form from plants and fungi), but higher quantities are synthesized in UV-B exposed skin. Irrespective of its origin, vitamin D is hydroxylated in the liver and the kidney to its biologically most active metabolite $1,25(OH)_2D_3$, which acts as a high-affinity ligand to the nuclear receptor VDR

that are performed primarily in the liver and the kidneys, respectively, create the biologically most potent vitamin D metabolite $1,25(OH)_2D_3$. As an example of a metabolized nutritional compound (see pathway B in Fig. 4.1), $1,25(OH)_2D_3$ acts as a specific ligand of the nuclear receptor VDR (Sect. 3.5). VDR is a transcription factor and regulates more than 1000 genes that are involved in the control of calcium homeostasis, bone formation, innate and adaptive immunity and cellular growth.

The outer surface of humans, the skin, mediates many interactions with the environment, such as thermoregulation, tactile sensitivity, detection of pain and the absorption of UV radiation. Besides these physiological functions, the color of skin, hairs and eyes represent the most obvious differences between human populations. People that live close to the equator or at high altitude, such as in the Himalayas or the Andes, have the darkest skin, while on the northern hemisphere at higher latitude lighter skin types are observed (Fig. 4.3). Some 100,000 years ago the skin of the anatomically modern humans was dark, because 1 million years earlier their ancestors turned dark when they lost most of their body hair, in order to better regulate their body temperature via sweating during endurance physical activity. Permanent dark skin better protects against the deleterious effects of solar UV-B

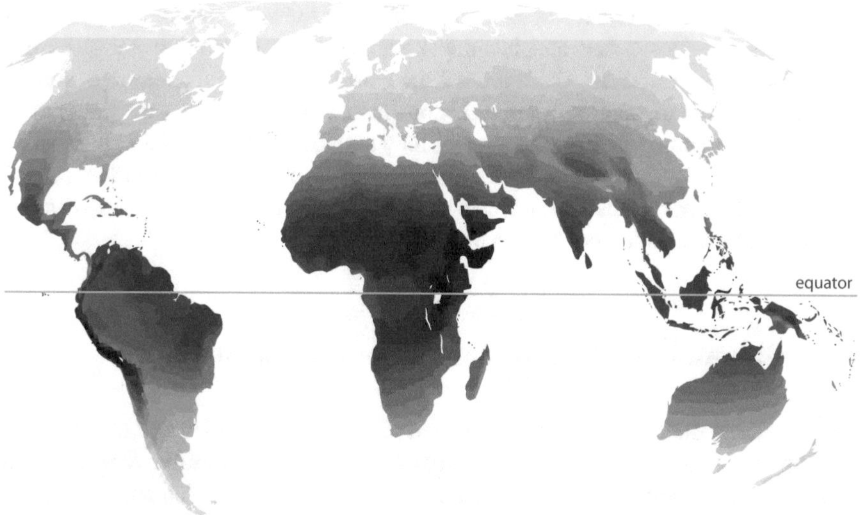

Fig. 4.3 World map of human skin color. The skin of original human populations is darker where UV-B radiation is strongest. This is close to the equator, at high altitude and by the oceans, as shown by the shading of the map

radiation, such as sunburns and skin cancer, but the main evolutionary driver for skin darkening most likely was the protection of the photosensitive B vitamin folate.

More than 30 genes were found to influence the pigmentation of human skin. In the process of melanogenesis the amino acids phenylalanine, tyrosine and cysteine are converted to melanin. SNPs in the genes encoding for the enzymes of this pathway as well as for ion channels in melanocytes or transporting molecules involved in melanosome maturation and export can result in light hair, light skin and blue eyes, as it is typical in northern European populations, or in dark skin and hair and brown eyes in equatorial population. The analysis of the haplotypes of these genes allowed estimating the time period when European skin became lighter. Changes within the *KITLG* gene that encodes for a ligand of the receptor tyrosine kinase KIT appeared already some 30,000 years ago, while the genes for the tyrosine converting enzyme TYRP1 and the ion transporters SLC24A5 and SLC45A2 were affected some 11,000–19,000 years ago. Interestingly, Europeans, Asians and first Neanderthals evolved light skin independently from each other. The evolutionary driver for the skin lightening genetic variations was most likely the low vitamin D_3 synthesis rate of darker skin in northern latitudes, which results in vitamin D_3 deficiency and serious consequences for evolutionary fitness, such as reduced bone strength and a defective immune system. For example, the tuberculosis rate of dark skin people living distant from the equatorial regions is significantly higher than that of individuals with light skin.

The correlation between human skin color evolution and the essential need for vitamin D_3 synthesis is only one of multiple examples how dietary needs and changes have directed human evolution over the past 60,000 years. In fact, radical changes in diet seem to have been a major driver of human evolution and probably even the main factor that enabled the modern human to survive and progress. Humans have spread from Africa around the world, experienced an ice age, domesticated hundreds of plant species and more than a dozen animals for the start of agriculture and dairy farming. Moreover, they significantly increased in population density and finally were exposed to a larger number of infectious diseases. In fact, infectious diseases, in particular malaria, have been among the strongest selective pressures in the most recent human evolution (Chap. 7).

4.3 Human Genetic Adaption to Dietary Changes

Humans were the only species that learned up to 1 million years ago to use fire to cook raw food and thereby created a safer and more easily digestible diet. Together with the use of tools and the omnivorous choice of diet, the advantage of cooking increased the energy yield for metabolism and allowed the enlargement of human brains. Interestingly, in today's human diet starch from grain flour, rice or potatoes is a major component, while early humans were not able to efficiently extract the energy from tubers and other starchy plant parts. The salivary enzyme amylase, encoded by a cluster of *AMY1* genes on chromosome 1, is the central human protein that hydrolyzes starch. The number of copies of the *AMY1* gene varies among human individuals (it can be up to nine copies per haploid genome) and directly correlates with enzyme concentrations. Interestingly, the genomes of other primates and that of Neanderthals carry only one copy of *AMY1* (as only 1–2 % of present humans). The average *AMY1* copy number within a human population relates to the starch content of their traditional diet, i.e. cultural variation in human diet explains some of the adaptive genetic differences between human populations. For example, in Japan large amounts of rice and starch from other sources are consumed reflecting many copies of the *AMY1* gene, whereas in the genetically closely related Siberian Yakut population (primarily eating fish and meat) significantly less *AMY1* copies are found. Since starch consumption is a central feature of agricultural societies, there seemed to be selective evolutionary pressure on the *AMY1* gene.

The genes of alcohol dehydrogenase (*ADH*) cluster that encode for ethanol metabolizing enzymes are another example of a gene locus that was positively selected when agriculture made the production of fermented alcoholic beverages easy. These examples suggest that the transition to new diet sources after the advent of agriculture and the colonization of new habitats seems to have been a major factor for the selection of human genes. Additional examples of genes that were positively selected due to dietary changes are: ADAM metallopeptidase with thrombospondin motif 19 and 20 (*ADAMTS19* and *ADAMTS20*), N-acylaminoacyl-peptide hydrolase (*APEH*), plasminogen activator, urokinase (*PLAU*) and ubiquitin protein ligase E3 component n-recognin 1 (*UBR1*), which encode for enzymes related to protein metabolism. Human population-specific examples are genes involved in metabolizing mannose (*MAN2A1* in West Africa and East Asia), sucrose

(*SI* in East Asia) and fatty acids (*SLC27A4* and *PPARD* in Europe, *SLC25A20* in East Asia, *NCOA1* in West Africa and leptin receptor (*LEPR*) in East Asia). Taken together, human populations have genetically adapted to their traditional diet, in order to make the most of local resources.

A prominent illustrative example is the co-evolution of dairy farming and the use of milk from infanthood to adulthood, referred to as lactase persistence. Lactase is an enzyme that is expressed in the intestine and splits the disaccharide lactose ("milk sugar") into glucose and galactose. Lactase non-persistence, also referred to as adult-type hypolactasia or lactose intolerance, represents an autosomal recessive trait that is characterized by diminished expression of the *LCT* gene after weaning. Lactose intolerance is the default genetic setup of early humans (as well as in most other mammals), maybe in order to avoid competition for mother milk between newborns and older children or even adults. It is still present in more than half of the human population, since the new variant of the *LCT* gene has emerged only within the last 10,000 years, primarily in parts of Europe after the start of dairy farming. Milk drinking created one of the strongest presently known selection pressures on the human genome that drove alleles for lactose tolerance to high frequency. Lactose tolerance is found also in livestock raising populations from Africa and Western Asia, but is almost completely absent elsewhere. Since milk is a perfect source of carbohydrates (primarily lactose), fat and calcium, the ability to use it as a reliable dietary source provides an enormous advantage for survival.

The SNPs associated with lactase persistence are located approximately 14 kb and 22 kb upstream of the *LCT* gene within exons 13 and 9 of the minichromosome maintenance type 6 (*MCM6*) gene on chromosome 2 (Fig. 4.4). The T/T allele at position -13,910 relative to the TSS of the *LCT* gene binds the transcription factor OCT1 with higher affinity than the C/T or the C/C variant. Also the transcription factors GATA6, CDX2 and HNF4A and HNF3A were described to associate with this regulatory region. Further SNPs at positions -22,018, -14,010, -13,915 and -13,907 in different human populations are also associated with lactose tolerance but did not show to be functional. As expected, the respective SNPs in the Neanderthal genome indicated that ancient human populations were lactose intolerant. Furthermore, the larger genomic region around the *LCT* gene demonstrates significant difference in the size of the respective haplotype blocks between lactose tolerant Europeans (more than 1 Mb) and non-Europeans, reflecting strong selection for the lactase persistence allele in particular in the Northern European population.

4.4 Regulatory SNPs and Quantitative Traits

GWAS analysis has indicated that 90 % of trait-associated variants are located outside of protein-coding regions of the human genome (Sect. 2.5). The SNP 13,910 kb upstream of the *LCT* gene that is associated in the northern European population with the trait lactose tolerance (Sect. 4.3) is a master example of a regulatory

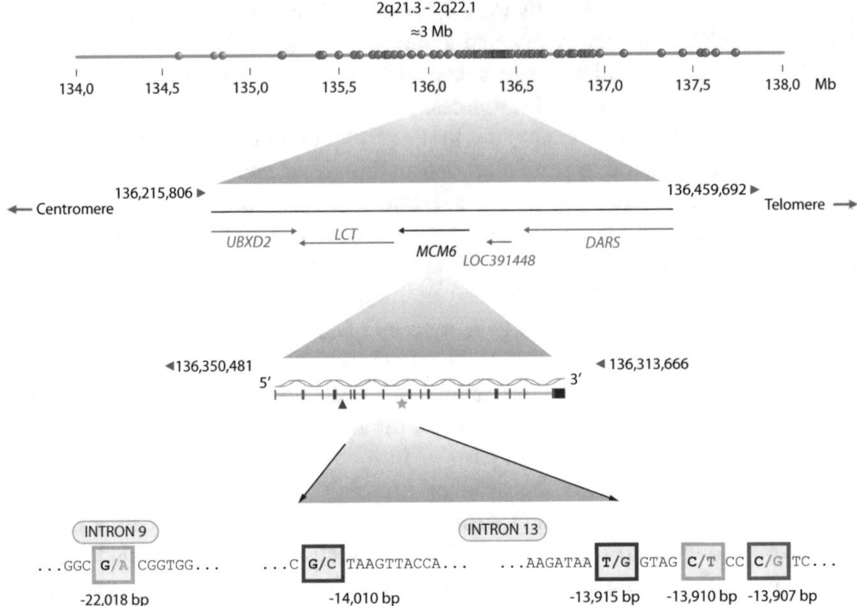

Fig. 4.4 Map of the genomic region of the *LCT* and *MCM6* gene. Location of the SNPs responsible for lactose tolerance within introns 9 and 13 of the *MCM6* gene in African and European populations

SNP. Regulatory variants seem to be equally important to SNPs affecting a protein-coding region in determining disease risks and traits (Fig. 4.5). The functional characterization of regulatory SNPs, such as the identification of transcription factor binding to the variant genomic region, may also suggest possible therapeutical interventions, for example, when the respective transcription factor is "drugable", i.e. when there is a synthetic or natural compound that modulates its activity. Gene regulatory events that are related to regulatory SNPs do not only depend on the sequence of the respective genomic site but also on its accessibility within chromatin. This emphasizes the impact of epigenomics on regulatory variations.

In contrast to the genome, which is identical in all 400 tissues and cell types of a human individual, the expression of genes depends on the tissue and the signals that it is exposed to, i.e. it represents the dynamic state of the cell. In the past, genome-wide gene expression was assessed via microarrays, but since a few years the next-generation sequencing method RNA-seq became the new standard for transcriptome profiling. In addition to mRNA expression, the method can monitor gene fusions, alternative spliced transcripts and post-transcriptional changes as well as the large variety of non-coding RNAs. The method expression quantitative trait locus (eQTL) mapping allows the functional genomic assessment of a genetic variant on the level of RNA. The principle of the eQTL approach is illustrated in Fig. 4.6.

Functional genetic variations can modulate various steps in the gene expression process from genes via mRNA to functional proteins. These are (i) changes the

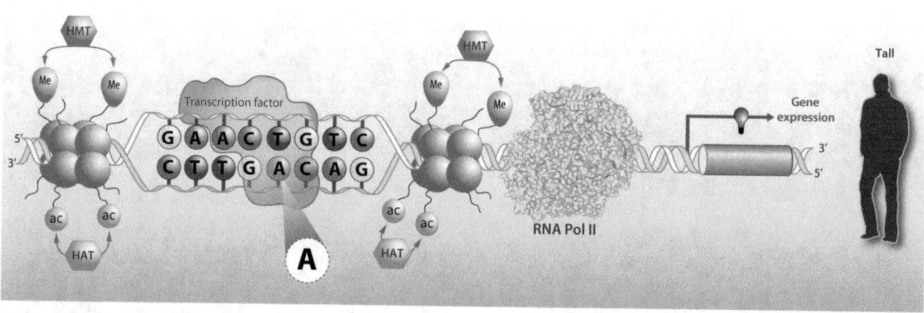

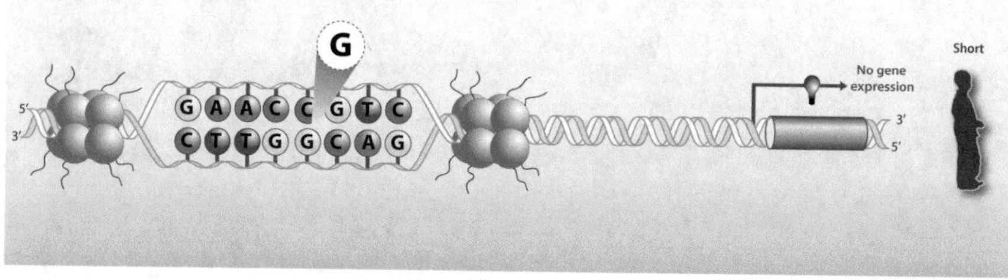

Fig. 4.5 The basis of human trait variation. Small variations within the DNA binding site for a transcription factor can faciliate and even enhance the association of this protein, such as the A (top) or inhibit its binding, when it is a G (*bottom*). The binding of the transcription factor influences the local chromatin structure via the activation of chromatin modifying enzymes, such as the histone acetyltransferases (HATs) and/or histone methyltransferases (HMTs), eventually leading to the activation of RNA polymerase II (Pol II) and the transcription of the respective gene. This may have a positive effect on the trait of interest, such as body height. In contrast, when the transcription factor does not bind, the respective genomic region remains inactive and the gene is not transcribed. This may have a negative effect on the studied trait

affinity of transcription factors for their genomic binding sites within promoters and silencers, (ii) a disruption of chromatin interactions, (iii) the modulation of the functionality of non-coding RNAs, (iv) the induction of alternative splicing and (v) an alternation in the post-translational modification pattern of proteins. The effect size of the functional SNPs can vary a lot and depend on the affected regulatory process and its epigenomic context, i.e. they are difficult to predict. At present, changes in the association of transcription factors due to variants in their specific binding sites are the best-understood types of regulatory SNPs (see examples in Box 4.1).

Box 4.1 Regulatory Variants with Impact on Obesity and CVD
Millions of DNA binding sites for the approximately 1600 DNA-binding transcription factors are encoded by the human genome. Dependent on their function and position, the regions, where these transcription factor-binding sites are clustering, are called promoters, enhancers, silencers or insulators. **A.** Obesity-associated variants in the *FTO* gene are rather distant to the gene's TSS. They were identified by the method of chromatin conformation capture

(continued)

Box 4.1 (continued)
(3C), in which DNA looping contacts between TSS and enhancer regions are detected. Interestingly, at least in mice, the same region also controls the brain-specific gene iroquois homeobox 3 (*Irx3*) in more than 500 kb distance. Interestingly, *Irx3* knockout mice showed an up to 30 % reduction in body weight through loss of fat mass and increased basal metabolic rate, suggesting that the IRX3 protein is involved in regulating body weight. **B.** Myocardial infarction and plasma levels of LDL-cholesterol were found to be strongly associated with genetic variants in chromosome 1. Fine mapping identified a SNP (rs12740374), for which eQTL analysis indicated most significant association with the gene sortilin 1 (*SORT1*). The more active minor allele created a binding site for the transcription factor CCAAT/enhancer binding protein (CEBP), i.e. it showed to be a gain-of-function regulatory SNP. More recent genomic approaches confirmed *SORT1* as a novel lipid-regulating gene and its pathway as a target for potential therapeutic interventions.

Results of the *ENCODE Project* have demonstrated that transcription factor binding sites regulating a given gene show a Gaussian distribution in relation to the the gene's TSS, i.e. they are found equally likely up- and downstream. Moreover, *ENCODE* data have created a unique genome-wide resource on (i) post-translational histone modifications, such as methylations and acetylations at various positions of the histones H3 and H4, indicating active and repressed enhancer and promoter regions, (ii) the chromatin accessibility, (iii) the genomic association of more than 100 transcription factors, (iv) DNA methylation indicating inactive genomic binding profiles and (v) the non-coding transcriptome in more than 100 human cell lines. Figure 4.6 provides an example how *ENCODE* data can be used for the functional characterization of a regulatory SNP. The set of annotations of the human genome are further extending for primary human tissues and cell types by data from the *FANTOM5 Project* and the *International Human Epigenome Consortium* (Box 2.4). The database *RegulomeDB* (http://regulomedb.org) combined information on chromatin state, transcriptional regulator binding and eQTLs and allows the identification and interpretation of functional DNA elements at the sites of regulatory variants.

4.5 "Omics" Analysis in Nutrition Sciences

From the perspective of nutrigenomics, food is a collection of dietary signals that are detected by cellular sensors, such as membrane proteins and nuclear receptors (Chap. 3), leading to changes in the transcriptome, proteome and metabolome, i.e.

Fig. 4.6 (continued) in suited tissues, in relation to the genotypes of the concerned SNPs (*center*). This approach demonstrated that the expression of gene 3 is significantly associated with the genetic variant. The use of genome annotation data from the *ENCODE Project* or comparable sources, such as histone modifications, transcription factor binding or accessible chromatin, may allow a mechanistic explanation of the function of the regulatory SNP (*bottom*)

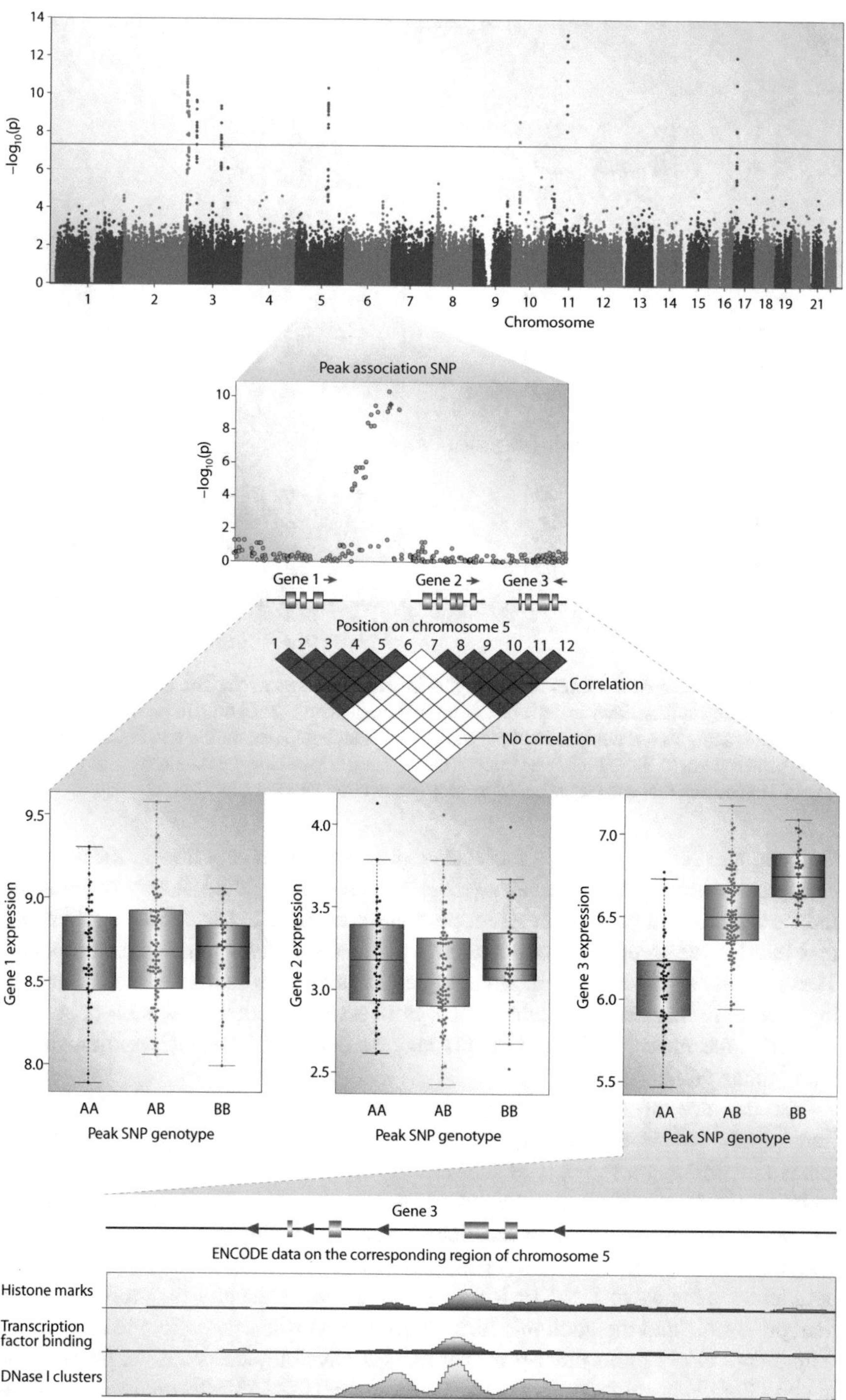

Fig. 4.6 Identifying regulatory SNPs by eQTL mapping combined with *ENCODE* data inference. The Manhattan plot representation (*top*) of a hypothetical GWAS study indicated a number of SNPs within a region of chromosome 5 that are with high statistical significance associated with the studied trait. The horizontal red line represents the p-value threshold. The regional association plot suggests three genes to be tested for eQTL mapping, such as mRNA expression,

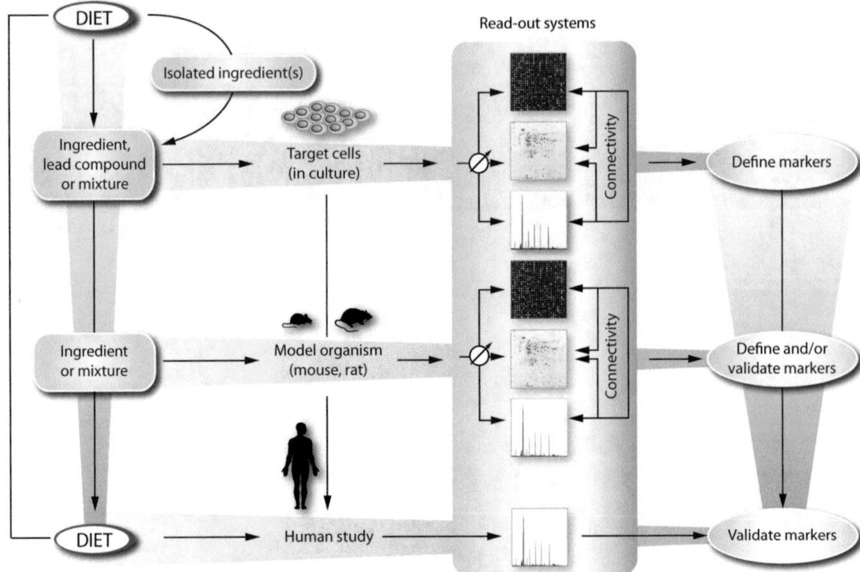

Fig. 4.7 Applications of omics technologies in nutrition research. The effect of food or its ingredients are studied either in cell culture, animal models or, e.g., in human intervention studies. In all cases, samples are analyzed via the use of omics technologies on the level of the transcriptome, proteome or metabolome. The integration of these large-scale datasets results in the definition of biomarkers that can be validated via complementary studies

the complete set of all mRNA molecules, proteins and metabolites in a cell or in a biological sample. Thus, food as a whole as well as individual dietary components induce a pattern in gene expression, protein expression and metabolite production that can be interpreted as "signatures" of the respective nutritional compound. These dietary signatures can be investigated *in vitro*, such as in cell lines representing metabolic organs, or *in vivo*, such as in rodent model organisms (Fig. 4.7). However, the most meaningful results may be obtained from intervention studies with human subjects.

Via the use of omics technologies, such as microarrays or RNA-seq for transcriptome-wide monitoring of mRNAs, nutrigenomics seeks to identify the genes that influence the risk of diet-related diseases. These genes are related to metabolic and/or regulatory pathways eventually providing a molecular explanation for the mechanism of action of the dietary compound. Nutrigenomics technologies that are based on the use of next-generation sequencing methods, i.e. the massive parallel sequencing of DNA or RNA molecules, are presently far more advanced than proteomic and metabolomic technologies. Nevertheless, proteome and metabolome data allow more precise measures of a physiological state. At present, the method liquid chromatography mass spectrometry (LC-MS/MS) analysis identifies up to 5000 of the most abundantly expressed proteins. The analysis of metabolites is performed either targeted, via gas chromatography-mass spectrometry (GC-MS) for a few hundred metabolites, or untargeted via LC-MS, which allows the detection of several thousand different molecules. A central point in nutrigenomic methodology is the integration of the data that were obtained from transcriptomic, proteomic and metabolomic profiling with specific nutrients or diets. Ideally, this results in the identification of biomarkers that can serve as early warning for nutrient-induced changes to homeostasis, such as the development of pre-diabetes (Chap. 10). In this way, nutrigenomic investigations can provide basic insight on the interaction between nutrition and the genome but can also serves as a diagnostic tool.

4.6 Integrative Personal Omics Profile

A proof-of-principle investigation demonstrating nowadays potential of next-generation technologies was provided by the iPOP analysis of one human individual. The iPOP study included a whole genome sequencing and sampling, carried out more than 20 times over a period of 14 months for (i) mRNA and miRNA expression in PBMCs (Box 2.2), (ii) proteome profile in PBMCs and in serum and iii) the metabolome and auto-antibodyome in blood plasma (Fig. 4.8). These molecular

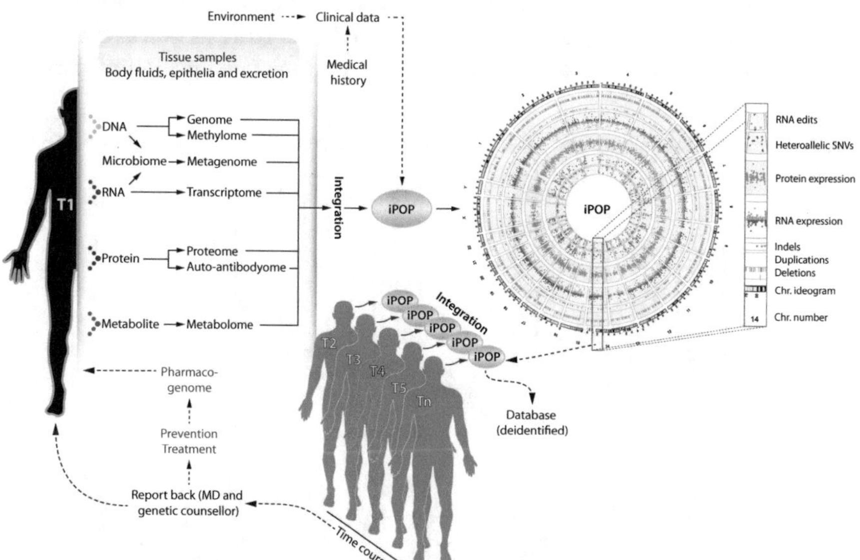

Fig. 4.8 Implementation of iPOP for personalized medicine. Tissue samples (for example, PBMCs) of an iPOP participant are collected at time points T1 to Tn, while diet, exercise, medical history and present clinical data are also recorded. The results of the iPOP analysis can be monitored by Circos plots (*right*), in which DNA (*outer ring*), RNA (*middle ring*) and protein (*inner ring*) data match to chromosome position. The data may be reported back to genetic counselors and/or medical practitioners, in order to allow most rational choices for prevention and/or treatment, which may be matched with pharmacogenetic data

datasets were complemented by medical lab tests for regular blood biomarkers. Interestingly, the frequent sampling enabled the detection of personalized physiological state changes, such as markedly elevated glucose levels in response following two viral infections during the investigation period. The integrative profile monitored both gradual trend changes as well as spike changes in particular at the onset of each physiological state adjustment. Thus, the iPOP analysis allowed a most comprehensive view on the biological pathways that changed during T2D onset of the study subject including dynamic changes in allele-specific expression and RNA editing events. Importantly, the T2D outbreak was detected in a very early stage, so that it could be effectively controlled and reversed by changing diet and intensified physical exercise of the individual.

The central aim of an iPOP-type analysis is the accurate assessment of disease risk of the investigated human individual. Due to the large number of genetic variants and the fact that many diseases are based on the combination of genetic and environmental factors, this aim is challenging. Results from genotyping approaches can be summarized in a "riskogram" that takes age, gender and ethnicity as well as multiple independent disease-associated SNPs into account, in order to calculate the subject's likelihood of developing a disease. The original iPOP study calculated for the investigated subject a previously unexpected increased risk to develop T2D (Fig. 4.9), which in fact was confirmed experimentally after a viral infection.

The field of personalized medicine significantly advanced due to the rapid development of omics technologies. Future personalized health care as well as the emerging field of personalized nutrition will benefit from the combination of personal

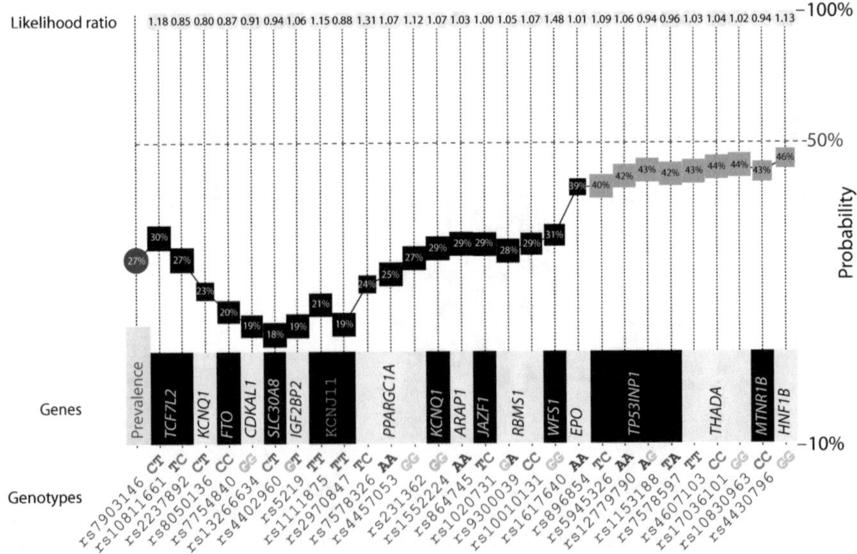

Fig. 4.9 The potential of personalized medicine. The "riskogram" illustrates how the volunteer's post-test probability of T2D was calculated on the basis of 28 independent SNPs. On *top*, the likelihood ratio is indicated, the central graph displays the post-test probability, while the associated genes, SNPs and the subject's genotypes are shown on the *bottom*

genomic information with global monitoring of the molecular profile that represent physiological states, as exemplified by the iPOP approach. iPOP-type investigations can be tailored and applied to monitor any disease or physiological state changes of interest. The integrative profile is modular and allows the addition of further omics information, such as epigenome-wide data and the microbiome of skin, oropharynx, nasopharynx, stomach, intestinal mucosa or urine, respectively, as well as quantifiable environmental factors. In this way, iPOP-like analyses may become also central to nutrigenomics.

Future View

In a few years whole genome sequence information will be available for millions of human individuals. This will allow a deeper understanding of the processes of human evolution and the causes of patterns of genetic variations for all human populations. The rapid development of omics technologies in combination with decreasing costs will allow collecting iPOP-style large-scale datasets on many individuals, the integration of which will allow further exploring the relationship between human genetic variations and complex diseases and respective traits. In particular the systematic exploration of epigenomics (Chap. 5) will provide critical insights into disease susceptibility. The ability to stratify individuals according to their genotype will make clinical trials more efficient by enrolling a lower number of subjects with an anticipated larger effect when personalizing the intervention. Diseases, such as T2D, will be classified into sub-phenotypes based on the genotype and the dynamic reply of the human individual, for example in response to a personalized diet. Nutrigenomics will assist in obtaining a comprehensive insight on the molecular links between nutrition and the (epi)genome. This will allow using diet for preserving health and for an improved personalized therapy, most likely in combination with synthetic drugs, in case of disease.

Key Concepts

- Human diet is a complex mixture of micro- and macronutrients, some of which (i) have a direct effect on gene expression, (ii) after metabolism modulate the activity of a transcription factor or (iii) stimulate a signal transduction cascade that ends with the induction of a transcription factor.
- When human individuals are classified according to the interplay of their lifestyle, metabolic pathways and genetic variation, the molecular insight based on nutrigenomics studies can suggest tailored diets, referred to as personalized nutrition, for early therapeutic intervention.
- The evolutionary driver for the skin lightening genetic variations was most likely the low vitamin D_3 synthesis rate of darker skin in northern latitudes, which results in vitamin D_3 deficiency and serious consequences for evolutionary fitness, such as reduced bone strength and a defective immune system.
- The average copy number of the salivary amylase-encoding gene *AMY1* within a human population relates to the starch content of their traditional diet, i.e. cultural variation in human diet explains some of the adaptive genetic differences between human populations.

- Drinking milk obtained from domesticated animals created one of the strongest presently known selection pressures on the human genome that drove alleles for lactose tolerance to high frequency.
- A SNP 13,910 kb upstream of the *LCT* gene is associated in the Northern European population with the trait lactose tolerance and represents a master example of a regulatory SNP.
- Food as a whole, as well as individual dietary molecules, induce a pattern in gene expression, protein expression and metabolite production that represent "signatures" of the respective nutritional compounds or the whole diet.
- Integration of large-scale datasets in nutrigenomics may result in the identification of biomarkers that can serve as early warning for nutrient-induced changes to homeostasis, such as the development of pre-diabetes.
- The iPOP analysis of one human individual represents a proof-of-principle study demonstrating nowadays potential of next-generation technologies. It allowed a most comprehensive view on the biological pathways that changed during T2D onset of the study subject.
- Future personalized nutrition and health care will benefit from the combination of personal genomic information with global monitoring of the molecular that represents physiological states, as exemplified by the iPOP approach.

Additional Reading

Frazer KA, Murray SS, Schork NJ, Topol EJ (2009) Human genetic variation and its contribution to complex traits. Nat Rev Genet 10:241–251

Järvelä I, Torniainen S, Kolho KL (2009) Molecular genetics of human lactase deficiencies. Ann Med 41:568–575

Knight JC (2014) Approaches for establishing the function of regulatory genetic variants involved in disease. Genome Med 6:92

Laland KN, Odling-Smee J, Myles S (2010) How culture shaped the human genome: bringing genetics and the human sciences together. Nat Rev Genet 11:137–148

Li-Pook-Than J, Snyder M (2013) iPOP goes the world: integrated personalized Omics profiling and the road toward improved health care. Chem Biol 20:660–666

Scheinfeldt LB, Tishkoff SA (2013) Recent human adaptation: genomic approaches, interpretation and insights. Nat Rev Genet 14:692–702

Sturm RA, Duffy DL (2012) Human pigmentation genes under environmental selection. Genome Biol 13:248

Chapter 5
Nutritional Epigenomics

Abstract Not only the genome but also the epigenome stores heritable information. The genome is supposed to stay stable during lifetime, while the epigenome is very dynamic and modulated by environmental stimuli, such as dietary molecules. It is expected that the missing classical genetic heritability of the susceptibility for complex diseases and traits might be eventually explainable via epigenetics. For example, persons that are born with low birth weight have a high risk to develop T2D later in their life, suggesting that epigenetic programming during embryogenesis and the intra-uterine environment both contribute to the T2D risk. Moreover, the lifestyle of the human individual, i.e. primarily the daily choice of diet, seems to create a metabolic memory within the epigenome. This concept suggests that lifestyle changes, such as the use of personalized diet, increased physical activity and consecutively weight loss can have a beneficial effect on the epigenome and thus on the risk for suffering from the metabolic syndrome (Chap 12).

In this chapter, we will define different epigenetic mechanisms, such as post-translational histone modifications and DNA methylation, that process information provided by dietary molecules. We will learn that many chromatin-modifying enzymes are susceptible to changes in the levels of intermediary metabolites acting as co-substrates and co-factors and respond to changes in nutrient intake and metabolism. Via the understanding of pre-natal supplementation in mouse models, we will get insight into the concepts of epigenetic programming. This will lead us to the thrifty hypothesis, and we will discuss the different approaches of epigenetic epidemiology including the concept of an "epigenetic drift" during adult life.

Keywords Epigenome • Epigenetic programming • Chromatin • DNA methylation • Histone modifications • Chromatin modifying enzymes • Intermediary metabolism • Acetyl-CoA • NAD^+ • Folate • Methylenetetrahydrofolate reductase • Agouti mice • Thrifty hypothesis • Epigenetic epidemiology • Epigenetic drift

C. Carlberg et al., *Nutrigenomics*, DOI 10.1007/978-3-319-30415-1_5

5.1 Epigenetics Mechanisms

Chromatin, a three-dimensional complex of genomic DNA and nuclear proteins, is subdivided into less densely packed euchromatin being easily accessible to transcription factors and compact heterochromatin, respectively, representing a functionally repressed state (Fig. 5.1a). Epigenetics is the study of functionally relevant modifications of the whole chromatin that do not involve a change in the genome sequence itself. Epigenetic alternations may remain stable during cell divisions and can last for multiple generations. The best example is the process of cellular differentiation, where due to epigenetic changes formerly totipotent stem cells become the various pluripotent cell lines of the embryo, which in turn are the precursors of terminally differentiated cells. The main epigenetic

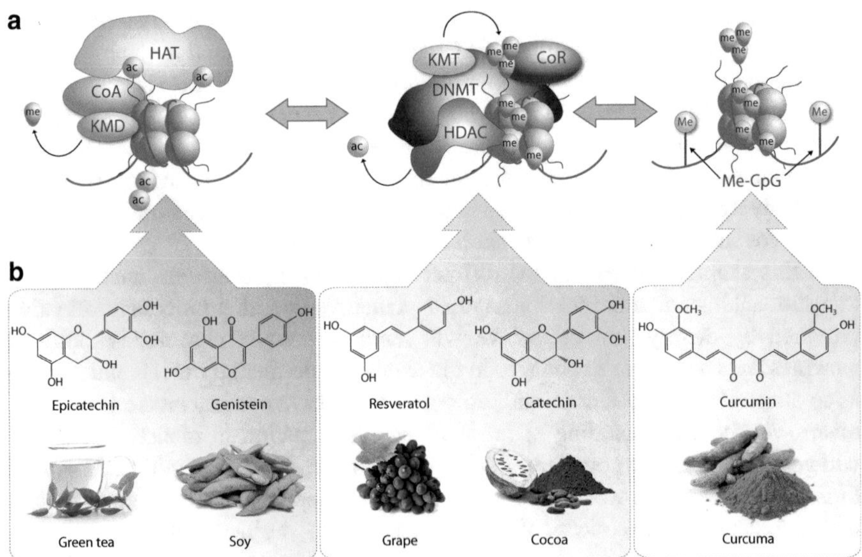

Fig. 5.1 Chromatin, associated proteins and nutrition-based epigenetic modulators. (a). Chromatin is distinguished into open chromatin (euchromatin, *left*) with loose nucleosome arrangement and closed chromatin represented by dense nucleosome packing (heterochromatin, *right*). There are several stages between these extremes that are summarized as facultative heterochromatin. Each stage is characterized by a set of chromatin modifying enzymes, such as HATs, HDACs, HMTs and histone demethylases (HMDs), DNA methyltransferases (DNMTs), co-activators (CoAs) and co-repressors (CoRs) of transcription factors that lead to the schematically indicated scenarios of acetylation and methylation of histone tails and genomic DNA. **(b).** The indicated plant-origin natural compounds have been shown to modulate the activity of chromatin modifying enzymes and in this way affect the epigenetic status of cells and tissues

mechanisms are (i) post-translational modifications of nucleosome-forming histone proteins (Box 5.1) and (ii) methylation of cytosines within genomic DNA (Box 5.2).

Box 5.1 Histone Proteins and Their Modifications

The basic, every 200 bp repeating unit of chromatin is the nucleosome, which consists of 147 bp genomic DNA that is wrapped nearly twice around an octamer of four pairs of the histone proteins H2A, H2B, H3 and H4. Histone proteins are rich in lysine residues, in particular at their amino-terminal tails that stick out from the nucleosome. The lysines but also arginine and serine residues are subject to post-translational, covalent, reversible chemical modifications, such as acetylations, methylations and phosphorylations, carried out by a variety of chromatin modifying enzymes. These histone modifications, referred to as epigenetic marks, represent a kind of chromatin indexing. Most marks are assigned to functional regions of the chromatin, such as promoters, enhancers or heterochromatin, the rules for which are summarized in the histone code. This code is a larger determinant for the accessibility of genomic binding sites of transcription factors and their associated co-factors, finally resulting in increased or decreased gene expression. Many chromatin modifiers and other nuclear proteins contain a set of common domains that specifically recognize different chromatin modifications, i.e. these proteins are able to "read" the histone code (Sect. 5.2).

Box 5.2 DNA Methylation

From the four DNA forming nucleotides only the one containing cytosine gets methylated and this in particular at the dinucleotide CpG, i.e. at sites where cytosine and guanine are found on the same DNA strand and are connected by a phosphodiester bond. CpG islands are genomic regions of at least 200 bp in length displaying a CG content of at least 55 % (compared to the 42 % average in the human genome). In the genome of normal human cells only 3-6 % of all cytosines are methylated, i.e. CpG islands are mostly unmethylated, and the genes in their vicinity keep their potential to be activated by transcription factors. DNMTs use S-adenosylmethionine (SAM) as a methyl group donor (Sect. 5.3) to methylate the carbon in 5′-position of cytosines. DNMT1 is a maintenance methyltransferase being active mainly during DNA replication showing a fidelity of 97-99.9 % per mitosis. In contrast, the enzymes DNMT3A and DNMT3B primarily function during embryogenesis and cellular differentiation and display a *de novo* methylation rate of 3-5 % per mitosis. The de-methylation of genomic DNA is regulated by ten-eleven translocation (TET) proteins, which convert 5-methylcytosine to 5-hydroxymethylcytosine, 5-formylcytosine and 5-carboxylcytosine. During DNA replication these modifications are then removed through iterative oxidation and base excision repair.

The *ENCODE Project* and other big biology consortia (Box 2.4) provide genome-wide maps of epigenetic marks in more than 100 human cell lines, primary cells and tissues. These data collectively indicate that in regions of open, active chromatin histone proteins are acetylated (for example, at lysine 14 of histone 3, H3K14ac) and genomic DNA remains unmethylated. In contrast, in repressed, closed chromatin histones are methylated (for example, at lysine 27 of histone 3, H3K27me) and also the DNA gets methylated. The change in DNA methylation during development starts with demethylation during cell devisions of the fertilized egg, followed by *de novo* methylation after implantation. Due to this epigenetic reprogramming during development, the intra-uterine period is considered critical for long-term health and disease risk (Sect. 5.5). However, DNA methylation does not only control the expression of specific genes during the development and differentiation of individual tissues, but it is also essential for silencing of imprinted genes, the second female X chromosome and retro-transposons (Box 5.3). Importantly, histone modifications precede or succeed DNA (de)methylations, i.e. both epigenetic processes display a cross-talk. For example, methyl-CpG binding proteins are capable of recruiting HDACs to methylated regions of genomic DNA.

Chromatin density plays an important role in regulating gene expression, primarily by controlling the accessibility of genomic binding sites for transcription factors. The dynamic competition between nucleosomes and transcription factors for critical binding sites is influenced by a large set of enzymes that either covalently modify histone proteins, termed chromatin modifiers, or move, reconfigure or eject nucleosomes, called chromatin remodelers. This process determines for each genomic region the density, composition and positioning of nucleosomes relative to the transcription factor binding sites that it contains.

A wide spectrum of secondary metabolites from fruits, vegetables, teas, spices, and traditional medicinal herbs, such as genistein, resveratrol, curcumin and polyphenols from green tea, coffee and cocoa, respectively, are able to modulate the activity of transcription factors and chromatin modifiers (Fig. 5.1b). Next-generation sequencing technologies allow the genome- and transcriptome-wide assessment of the specificity and efficacy of these compounds, for example, in order to prevent and treat cancer.

Box 5.3 Genomic Imprinting

Due to the phenomenon of genomic imprinting in diploid cells some genes are expressed in a parental-origin-specific manner, i.e. either the gene copy of the father or that of the mother is used. The imprinting process is essential for normal mammalian growth and development, such as exemplified by the inactivation of the second X chromosome of females during early embryogenesis. The regulatory region of the repressed copy of imprinted genes is hypermethylated and located within heterochromatin. This repression is inherited through mitotic division in somatic cells. Most (but not all) of the genomic imprinting is erased in primordial germ cells of the next generation and reset as they differentiate into the parental gametes.

Box 5.4 Levels of Gene Expression

Precise coordination of transcriptional programs allows cells to adapt to changes in environmental signals. Therefore, the mRNA expression of genes is tightly controlled on three different levels. The first level, the DNA code, is the sequence of the human genome, which contains some 200,000 protein-coding exons that have to be transcribed and combined by splicing into mature mRNA molecules. In addition, the human genome codes for some 20,000 non-coding RNAs, most of which are used for fine-tuning and feedback control of gene expression ranging from chromatin accessibility to protein translation. The second level, the epigenetic code, is controlled by a combination of the histone code (see Box 5.1) and the rules for DNA methylation (see Box 5.2). The third level, the transcription factor program, represents the relative amount and activity state of the transcription factors that are expressed in a given cell. In turn, transcription factor expression is controlled by both the DNA code and the epigenetic code. In addition, most transcription factors are regulated in their activity by a number of different processes, such as post-translational modification, ligand-binding, dimerization and translocation from the cytosol to the nucleus. Moreover, micro RNAs (miRNAs), which are the evolutionary oldest regulators of gene expression, mostly counter-act the activities of transcription factors and thus fine-tune the stability of mRNAs.

In general, gene expression is controlled on the level of (i) the genomic DNA sequence (the DNA code), (ii) the accessibility of genomic DNA (the epigenetic code) and (iii) the abundance and activity of transcription factors (the transcription factor program) (Box 5.4). Most trait-associated variations of genomic DNA are located outside of protein-coding regions, such as the major SNP controlling the persistence of *LCT* gene expression after weaning (Sect. 4.3). These regulatory SNPs offer the possibility to understand the mechanistic function of those SNPs that do not change the amino acid composition of a protein. However, since regulatory SNPs are a part of the DNA code, they do not close the gap in the missing classical genetic heritability. Thus, the latter may be explained via the epigenetic code or the transcription factor program.

5.2 Intermediary Metabolism and Epigenetic Signaling

Within a cell, signal transduction pathways result in the activation of gene expression programs (Box 5.4) that integrate environmental inputs, such as the availability of energy substrates, in order to mediate responses targeting homeostasis. Numerous connections between products of intermediary metabolism and chromatin modifying enzymes (Box 5.5) are known. The human genome tissue-specifically expresses hundreds of chromatin modifiers that interpret (“read”), add (“write”) or remove (“erase”) post-translational histone modifications. So-called “bromodomains” are found in all type of proteins that are able to recognize acetylated residues, such as

Box 5.5 Chromatin Modifiers
The activity of chromatin is modulated by a group of enzymes that catalyze rather minor changes in residues of histone proteins, such as the addition or removal of acetyl- or methyl-groups. Chromatin acetylation is generally associated with transcriptional activation, while the exact amino acid residue of the histone tails that is acetylated seems not to be very critical. The acetylation state of a given chromatin locus is controlled by two classes of antagonizing histone modifying enzymes, HATs and HDACs. In analogy, also for histone methylation there are two classes of enzymes with opposite functions, HMTs and HDMs. Although histone methylation mainly mediates chromatin repression, at certain residues, such as H3K4, it results in activation. Therefore, for histone methylation the exact residue in the histone tail and its degree of methylation (mono-, di- or tri-methylation) is of critical importance.

HATs, HMTs, chromatin remodeling enzymes, co-activators and general transcription factors. In contrast, "chromodomains" are far more specific for a given chromatin modification, i.e. chromodomain-containing nuclear proteins recognize their genomic targets with far more accuracy than bromodomain proteins. The activity of almost all of these chromatin-modifying enzymes critically depends on intracellular levels of essential metabolites, such as acetyl-CoA, uridine diphosphate (UDP)-glucose, α-ketoglutarate, NAD^+, flavin adenine dinucleotide (FAD), ATP or SAM. Since the cellular concentrations of several of these metabolites represent the metabolic status of the cell, the activities of the chromatin modifiers reflect the intermediary metabolism. The enzymes act as sensors and process the metabolic information into dynamic post-translational histone modifications, which coordinate adaptive transcriptional responses (Fig. 5.2). Therefore, such a sensing of metabolic signals modulates the activity of gene regulatory networks that in turn control cell fate decisions. Critical disturbances in energy metabolism may also lead to stable epigenetic changes that are maintained through the germ line and may affect the health of the next generations (Sect. 5.5).

Acetyl-CoA is generated when ingested nutrients enter the catabolic pathways of intermediary metabolism (Fig. 5.3). This suggests that histone acetylation via HATs primarily depends on cytosolic pools of acetyl-CoA. The acetylation of histones during feeding, i.e. under high-energy conditions, leads to transcriptionally open chromatin, thus allowing the expression of genes that regulate cellular proliferation, lipogenesis and adipocyte differentiation. In turn, histone deacetylation via HDACs is also connected to intracellular energy levels. For example, the beneficial effects of calorie restriction (Sect. 6.3) on metabolic parameters depend on sensing of the energy carrier NAD^+ by sirtuin-type HDACs. Interestingly, NAD^+ levels fluctuate in a circadian manner, which links, via epigenetic mechanisms involving sirtuins, the peripheral clock to the transcriptional regulation of metabolism (Sect. 3.6). Importantly, also non-histone proteins, such as the transcription factor p53 or the

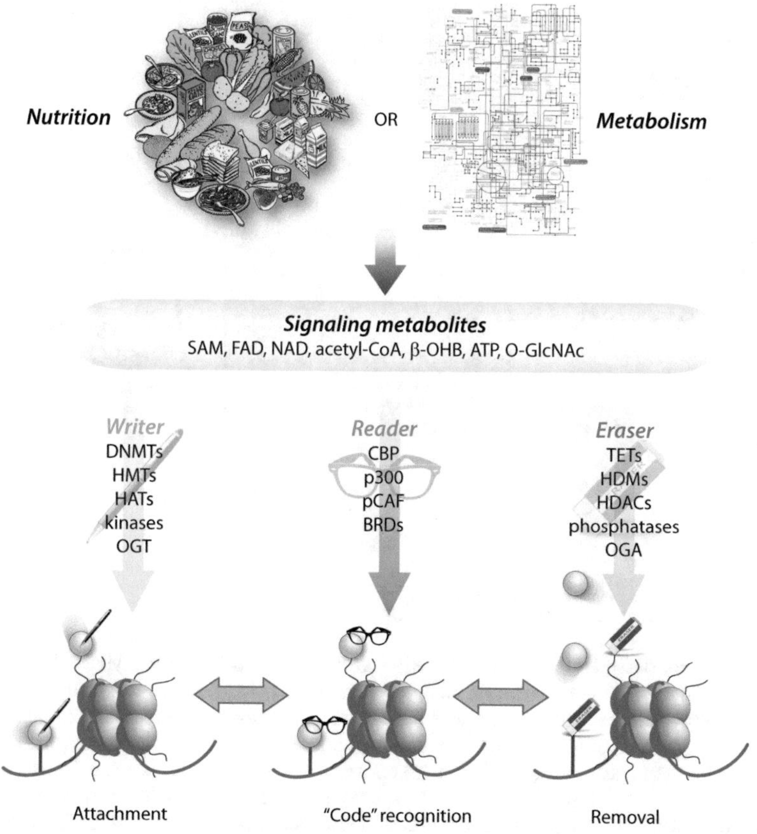

Fig. 5.2 Epigenetic mechanisms link metabolites and transcription. Changes in nutrition or fluctuations in metabolism affect the transcriptional responses of metabolic tissues. Several intermediary metabolites change the activity of chromatin-modifying enzymes in a dose-dependent manner. These proteins use some of these metabolites as co-substrates and/or co-factors and act in this way as metabolic sensors. "Writer" enzymes create covalent chromatin marks, "reader" enzymes recognize these marks and "eraser" enzymes remove them. These histone tail modifications create changes in the local chromatin structure, which has consequences for the activity and regulation of the neighboring genes

co-activator peroxisome proliferator-activated receptor gamma, coactivator 1 alpha (PPARGC1A), are deacetylated by sirtuins (Sect. 6.4).

The short-chain fatty acid butyrate is a potent inhibitor of several HDACs and is produced from dietary fibres in the lumen of the colon. Fibre-rich diet can prevent colitis and colon cancer in humans, which could be, at least in part, explained via butyrate-mediated inhibition of HDAC-dependent transcriptional programs of colonocyte proliferation. Another link of the epigenetic code to intermediary metabolism is provided by the addition and removal of N-acetylglucosamine (O-GlcNAc) to serine and threonine residues of histones by the enzymes O-GlcNAc transferase (OGT) and O-GlcNAcase (OGA). The substrate of this glycosylation reaction, UDP-glucose, derives from the hexosamine pathway and reflects changes in glucose levels (Fig. 5.3).

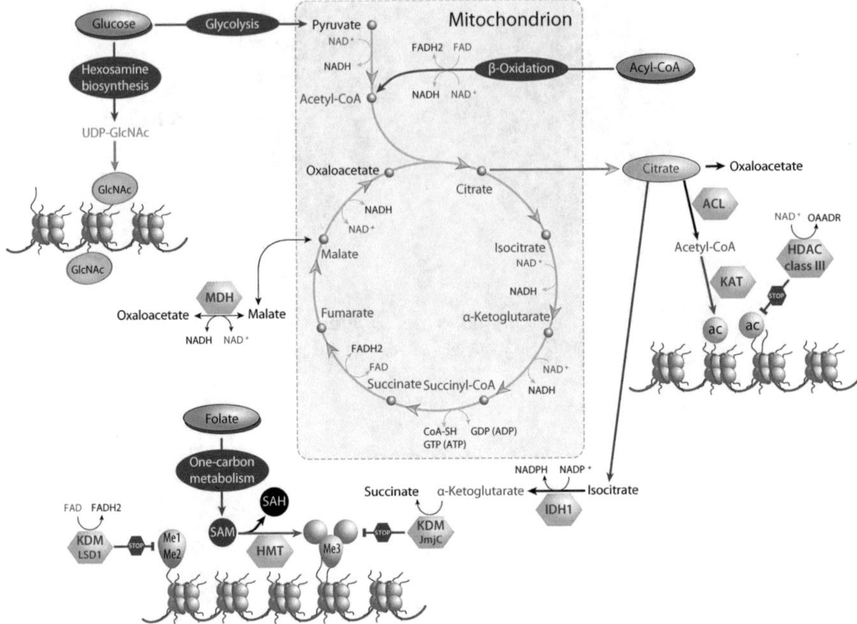

Fig. 5.3 Links between metabolic pathways and chromatin signaling. Metabolites that are essential for the function of chromatin modifiers participate in key biochemical pathways of intracellular energy balance. The TCA cycle is the central connection between catabolic and anabolic pathways. Glycolysis of glucose and β-oxidation of fatty acids are catabolic processes that generate acetyl-CoA, whereas during periods of glucose excess acetyl-CoA is removed from mitochondria via the citrate shuttle that fuels lipogenesis and the biosynthesis of various other macromolecules. This also provides acetyl-CoA for histone acetylation via HATs. Glucose is also used via the hexosamine biosynthetic pathway generating the co-enzyme UDP-GlcNAc that together with OGT mediates histone O-GlcNAcylation. Folate is a micronutrient that acts as a methyl-group donor for DNA and histone methylation. It is the central substrate of the one-carbon metabolism, i.e. of a cyclic reaction generating SAM (Sect. 5.3). Finally, the NAD^+/NADH ratio in mitochondria is connected with the malate-aspartate shuttle and controls the activity of sirtuins (Sect. 6.5)

5.3 One-Carbon Metabolism and DNA Methylation

Methyl donors are critical during pregnancy and dietary excess as well as deficiency may have an impact on epigenetic programming in mice (Sect. 5.4) and humans (Sect. 5.5). The methyl donor substrate SAM connects DNA methylation with intermediary metabolism. SAM is generated from the amino acid methionine and ATP in the one-carbon metabolism pathway (Fig. 5.4). When a methyl-group of SAM is transferred to DNA or a histone, the product S-adenosylhomocysteine (SAH) is recycled back to SAM. Interestingly, SAH acts as a negative feedback regulator for HMTs, suggesting that the SAM/SAH ratio, referred to as the "methylation index", is critical for histone and DNA methylation. A derivative of the B vitamin folate,

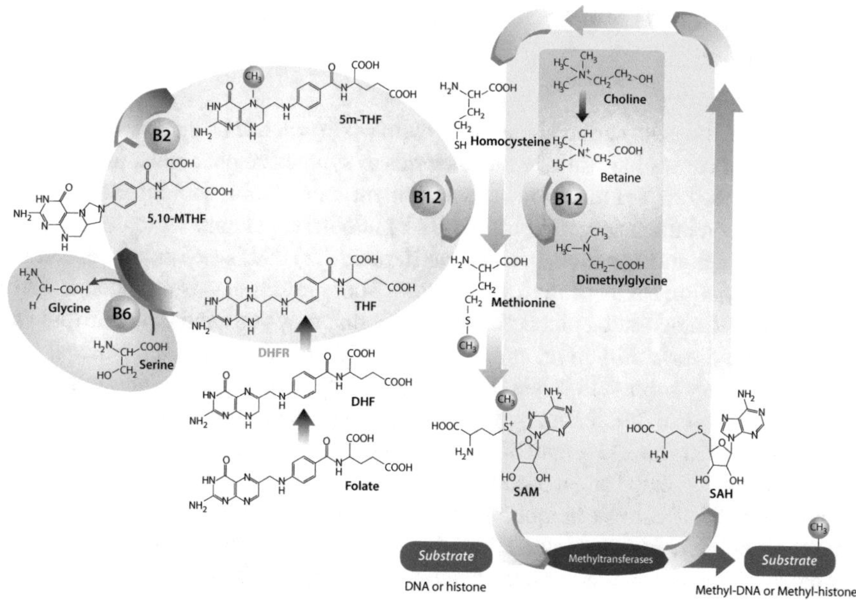

Fig. 5.4 Metabolism and methyltransferases. DNA and HMTs use the methyl-group of SAM and generate SAH that is converted to homocysteine. In a vitamin B12-dependent reaction that utilizes carbons derived from either choline or folate, homocysteine is converted back to methionine. Also shown are steps that require the vitamins B6 and B12. For simplicity, the respective converting enzymes are not shown

tetrahydrofolate, serves as a methyl-group donor for feeding of the cyclic one-carbon pathway. The dependence of the pathway on folate and other micronutrients is another example of the direct connection between nutrition and epigenetics. The function of folate in normal neural tube closure in early gestation (21-28 days after human conception) is well known and maternal supplementation with folate is recommended for prevention of neural tube defects.

Elevated homocysteine is a well-studied indicator of one-carbon metabolism disturbance and is related to low concentrations of folate, vitamins B12 and B6, choline and betaine. Moreover, a variant in the human *MTHFR* gene encoding for the methylenetetrahydrofolate reductase that catalyzes the conversion of 5,10-methylenetetrahydrofolate to 5-methyltetrahydrofolate (Fig. 5.4) reduces the activity of the enzyme by some 50 %. Homozygosity for the missense SNP C677T (rs1801133) causes the exchange of alanine at position 222 of the protein into valine. This genetic variation is observed in 10-15 % of Europeans. Low folate intake affects individuals with the TT genotype to a greater extent than those with the CC or CT allele. Accordingly, raised plasma homocysteine concentrations cause an elevated risk of pre-mature delivery, low birth weight and neural tube defects.

5.4 Nutrition-Triggered Transgenerational Epigenetics in Mice

In the mouse, the gene agouti signaling protein (*Asip*) encodes a paracrine signaling molecule that causes hair follicle melanocytes to synthesize pheomelanin, a yellow pigment, instead of the black or brown pigment, eumelanin. The insertion of the retrotransposon intracisternal A particle (IAP) into the regulatory region of the *Asip* gene creates a dominant allele of the gene (termed A^{vy}) that is expressed depending on the methylation status of the IAP element. Some retrotransposons, such as IAP, show so-called stochastic epigenetic variability that may substantially contribute to the phenotypic variability in mammals. In case of A^{vy}/a mice this results in coat colors that can vary from yellow via mottled to wild-type dark coat color, i.e. A^{vy} is a metastable epiallele. Interestingly, yellow coat mice become obese and are hyperinsulinemic. When the IAP element is methylated, the synthesis of the yellow pigment is down-regulated and a pseudoagouti dark coat color is produced. In contrast, unmethylated IAP allows ubiquitous *Asip* expression leading to both yellow coat color and obesity. Another interesting characteristic of the A^{vy} mouse model is its capacity for epigenetic inheritance, meaning that, when A^{vy}/a animals inherit the A^{vy} allele maternally, *Asip* expression and coat color correlate with the maternal phenotype. In addition, the mottled phenotype indicates that IAP methylation is mosaic, i.e. the *Asip* gene is not expressed in all cells. This suggests that the methylation pattern of the IAP element is established early in development and the coat color provides an easy phenotypic readout of the methylation and expression status of the element throughout life. All these properties make the A^{vy} *Asip* mouse an ideal *in vivo* model, in order to investigate a mechanistic link between environmental stimuli, such as nutrition, and epigenetic states of the genome (for details see below).

This was supported by the following experiment: Two weeks before mating with male A^{vy}/a mice, female wild-type a/a mice were either supplemented or not with methyl donors, such as folate, vitamin B12 and betaine (Fig. 5.5). The supplementation was continued during pregnancy and lactation. While the F1 generation of non-supplemented mothers displayed the expected number of yellow color phenotypes, the offspring of supplemented mothers shifted toward a brown coat color phenotype. This suggests that maternal methyl donor supplementation leads to increased A^{vy} methylation in the offspring. The inheritance of an epigenetic programming to the next generation indicates that at least metastable epialleles, such as IAP, are able to resist the global demethylation of the genome before preimplantation. Interestingly, when A^{vy} mice were fed with a soy polyphenol diet causing changes in their DNA methylation patterns, their offspring was protected against diabetes, obesity and cancer across multiple generations.

In a mouse model of maternal undernutrition that leads to low birth weight and glucose intolerance in male and female F1 offspring, it could be shown that exposure to suboptimal nutrition during fetal development leads to changes in the germ-cell DNA methylome of male offspring even when these males were nourished normally after weaning. These phenotypic differences are transmitted through the

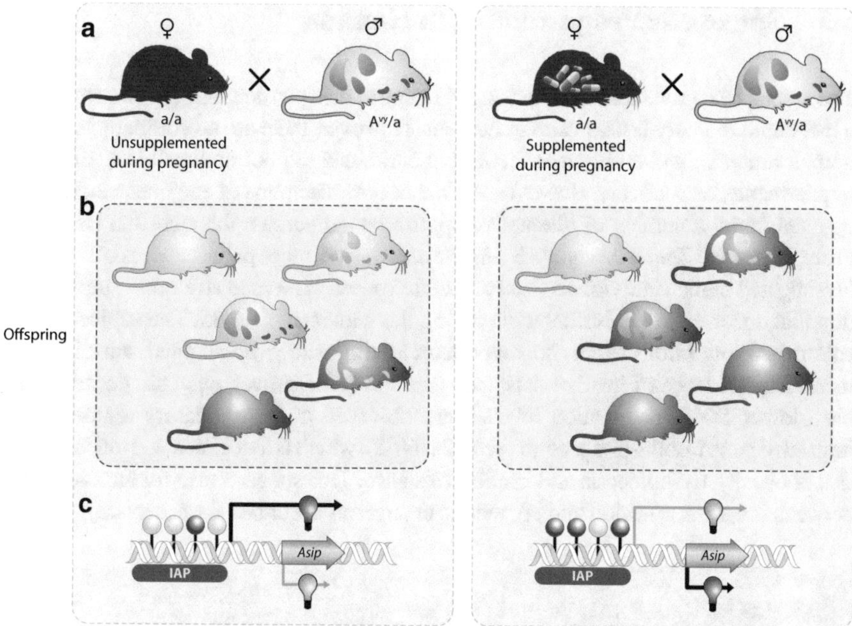

Fig. 5.5 Maternal dietary supplementation affects the phenotype and epigenome of A^{vy}/a offspring. (**a**). The diets of female wild-type a/a mice are either not supplemented (*left*) or supplemented with methyl-donating compounds (*right*), such as folate, choline, vitamin B12 and betaine, for 2 weeks before mating with male A^{vy}/a mice, and ongoing during pregnancy and lactation. (**b**). The coat color of offspring that are born to unsupplemented mothers is predominantly yellow, whereas it is mainly brown in the offspring from mothers that were supplemented with methyl-donating substances. About half of the offspring from these matings do not contain an A^{vy} allele, are therefore black (a/a) and are not shown here. (**c**). The molecular explanation of DNA methylation and *Asip* gene expression. Maternal hypermethylation after dietary supplementation shifts the average coat-color distribution of the offspring to brown by causing the IAP element upstream of the *Asip* gene to be more methylated on average than in offspring that are born to mothers fed an unsupplemented diet. The arrow size is proportional to the amount of ectopic and developmental *Asip* gene expression. White circles indicate unmethylated CpGs and red circles are methylated CpGs

paternal line to the F2 offspring. In the genome-wide view of DNA methylation, more than 100 genomic regions in the F1 sperm from maternally undernourished male offspring were hypomethylated compared with the equivalent regions of sperm from control offspring. The presence of novel hypomethylated regions suggested that primordial germ cells from nutritionally restricted fetuses did not completely remethylate their DNA.

Taken together, the different rodent models indicate that epigenetic memory can be passed from one generation to another by inheriting the same indexing of chromatin marks. However, while short-term "day-to-day" responses of the epigenome are primarily mediated by non-inherited changes in the histone acetylation level, DNA methylation seems to be designed in particular for a long-term cell memory.

5.5 Epigenetic Programming in Humans

The results from rodent models (Sect. 5.4) impose the question whether the concept of a metabolic memory is also valid in humans. However, there are no comparable natural human mutants, and embryonal feeding experiments are not compatible with ethical requirements, respectively. However, the rather new discipline of epigenetic epidemiology has found a number of alternative approaches to address this question (Box 5.6). From these, the *Dutch Hunger Families Study* raised most public interest. The study investigated human individuals that had been exposed *in utero* to an extreme undernutrition that occurred in the Netherlands during the winter of 1944/45. The subjects in the cohort had low birth weight, however, later in life were overweighted and displayed increased incidence of insulin resistance (Sect. 9.4). Moreover, even six decades after birth lower DNA methylation levels were observed at the regulatory region of the imprinted gene insulin-like growth factor 2 (*IGF2*), which is associated with an increased risk of obesity, dyslipidemia and insulin resistance. This suggests also for humans a link between pre-natal nutrition and epigenetic changes as described for rodents.

Box 5.6 Methods of Epigenetic Epidemiology

Natural experiments: Studies in which the exposure to a condition is not under experimental control. For example, in the Netherlands the *Dutch Hunger Winter Families Study* (www.dutchfamine.nl) follows individuals who were exposed as an embryo to famine of their mothers during the winter of 1944/45.

Longitudinal birth cohorts: These studies analyze epigenetic changes over time and relate them to environmental exposures and the onset of disease. For example, in the UK the *Avon Longitudinal Study of Parents and Children* (www.bristol.ac.uk/alspac) follows the health of 14,500 families, who had children born in 1991/92.

Longitudinal twin studies: Since monozygotic twins have the same parents, birth date and sex, they are cross-wise the ideal references for the study of individual epigenetic variations. For example, in Australia the *Peri/Postnatal Epigenetic Twins Study* (www.mcri.edu.au/research/research-projects/pets) explores epigenetic variation between twins at birth.

Pre-natal cohorts: The pre-natal developmental period is considered to be the most crucial time frame for epigenetic programming. When individuals are tracked from the point of conception onwards, maternal and intra-uterine influences, their genotype and postnatal environment can be related to health and disease. For example, in the UK the *Southampton Women's Survey* (www.mrc.soton.ac.uk/sws) follows 12,500 women and their offspring through their pregnancies.

In vitro fertilization conception cohorts: The *Danish National IVF Cohort Study* studied frequency of imprinting diseases in children born after *in vitro* fertilization (IVF).

The time point and frame of the different types of epigenetic studies are visualized in Fig. 5.6.

(continued)

Box 5.6 (continued)

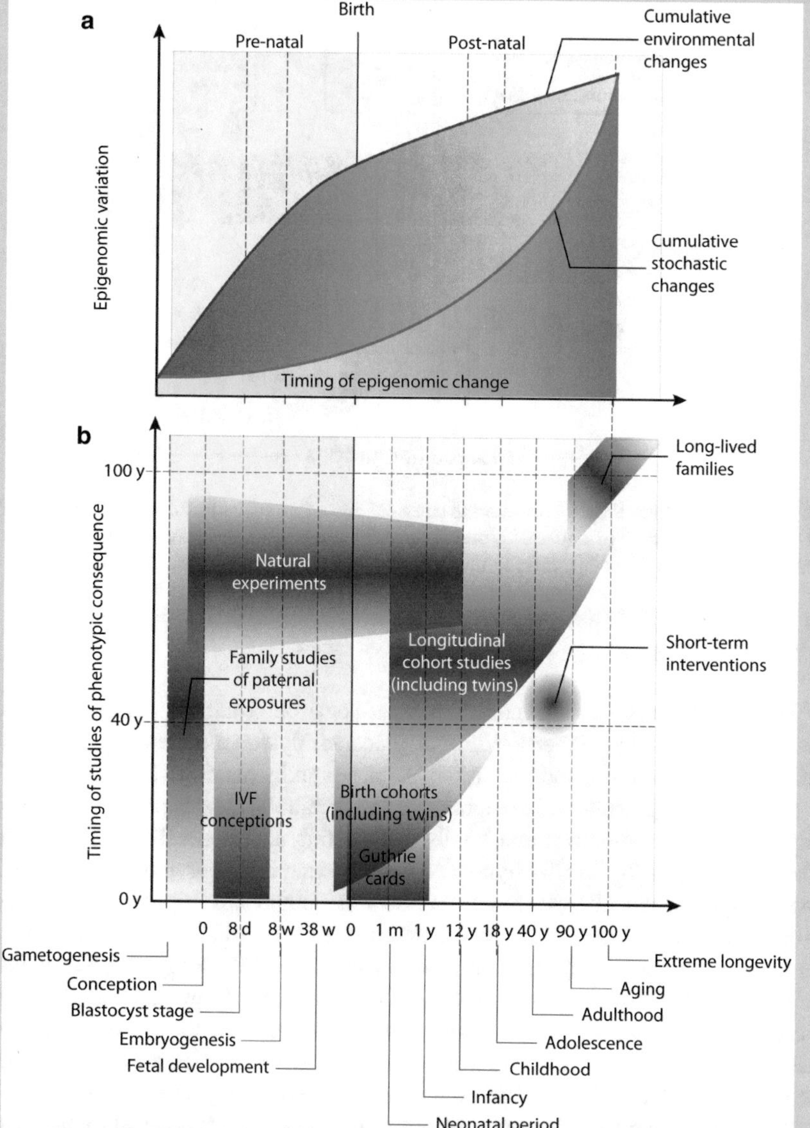

Fig. 5.6 Analyzing epigenetic variation in populations. (**a**). Epigenetic changes can occur at any time during life, but there is significantly increased sensitivity during early pre-natal development. (**b**). The pre-natal epigenetic changes may be investigated using IVF cohorts (Box 5.6), archived Guthrie cards, and birth cohorts tracking life from as early as periconception. Historical famines represent the few opportunities to link the pre-natal environment to health outcomes later in life. Longitudinal cohort studies (especially involving twins) sample peripheral tissues and take biopsies from disease-relevant tissues. Short-term (dietary) interventions can identify specific dietary compounds that induce tissue-specific epigenetic modifications, while long-lived families can help in identifying the importance of maintaining epigenetic control for healthy aging

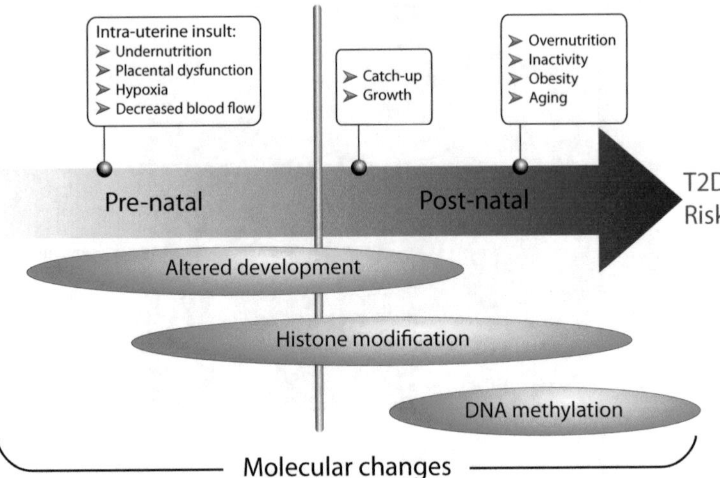

Fig. 5.7 Thrifty hypothesis. Intra-uterine stressors, including maternal undernutrition or placental dysfunction (leading to impaired blood flow, nutrient transport, or hypoxia) can initiate abnormal patterns of development, histone modification and DNA methylation. Additional post-natal environmental factors, including accelerated post-natal growth, obesity, inactivity and aging can further contribute to T2D risk, potentially via further histone modifications and DNA methylation in metabolic tissues

Human mothers being exposed to an adverse environment during embryonic development, such as undernutrition or placental dysfunction leading to impaired blood flow, nutrient transport or hypoxia, causes an increased risk for the offspring to develop during adulthood symptoms of the metabolic syndrome, such as obesity, impaired glucose tolerance and finally T2D during adulthood. These observations were the basis for the "thrifty hypothesis" suggesting that poor nutrition in early life produces permanent changes in glucose-insulin metabolism (Fig. 5.7). Fetal malnutrition leads to impaired fetal growth, and low birth weight favors a thrifty phenotype that is epigenetically programmed to use nutritional energy efficiently, i.e. to be prepared for a future environment with low resources during adult life. When these individuals are exposed to an obesogenic environment, they have an increased risk of obesity and its consequences on multiple diseases (Chap 8), i.e. the metabolic memory during fetal programming *in utero* has an impact in adult life.

The thrifty hypothesis has been extended to a more general concept of the "developmental origins of health and disease", suggesting that the early-life development is critically sensitive to inadequate nutrition. Other environmental factors in this context lead to permanent changes in metabolism that can alter susceptibility to complex diseases, such as diabetes, obesity, CVD, hypertension, asthma and cancer. It is obvious that epigenetic changes during embryonic development have a much greater impact on the overall epigenetic status of the organism than adult stem cells or somatic cells, since they are transmitted over far more consecutive mitotic divisions. These exposures in early life are recorded and stored as cellular memory through persistent adaptations in cellular functions with long-term effects. This means that the phenotype of a human individual is the result of complex genome-environment interactions leading to life-long remodeling of the epigenome.

The concept of epigenetic programming via nutritional compounds includes the idea that dietary interventions of human adults, such as calorie restriction (Sect. 6.3), Mediterranean or Nordic diet, all can affect the chromatin status of the subjects and lead to the expression of genes being beneficial to metabolic health. This implies that epigenetic states initially fixed during embryogenesis may shift in response to intrinsic and environmental factors, such as nutritional compounds. This epigenetic drift of metastable epialleles has been described, for example, in monozygotic twins that start with an indistinguishable epigenome but drift to different stages during their life. Changes in diet or in metabolism being associated with obesity may cause an epigenetic drift that may be inherited to the following generations. If this concept holds true, the worldwide growing epidemic of obesity and metabolic disease may lead to a significant epigenetic predisposition for the metabolic syndrome in the subsequent generations resulting in a vicious cycle (Fig. 5.8). The idea of an inherited genetic drift includes that epigenetic modifications acquired during chronological aging reduce the capacity for homeostatic responses to nutritional stress, such as overnutrition, leading to an overall decrease in life expectancy.

Nutrigenomics aims on the genome-wide understanding of the effects of nutrients and their metabolites on intermediary metabolism and its influence on the epigenome. Ideally, this should not only include easily accessible tissues, such as blood, but also metabolic organs, such as liver, pancreas, adipose tissue and muscle. For example, the DNA methylome of pancreatic islets from individuals with T2D and those from healthy individuals showed differences at more than 800 gene loci. Importantly, some of these methylation changes correlated with mRNA levels suggesting differences in β cell function. In this way, nutrigenomic approaches will maintain wellbeing, promote health and open up new therapeutic strategies, such as a possible reprogramming of the epigenome of metabolic organs through personalized diet including natural compounds that modulate the activity of chromatin modifiers and transcription factors (Sect. 5.1).

Future View

For future epigenetic studies the choice of tissue type will be a critical issue. Epigenome mapping may be preferentially performed with easily accessible cells from whole blood, buccal mucous membrane or intestines (harvested via stool). Comparisons of epigenetic marks within different tissues of the same individual will create an inventory of epigenetic marks that may be extrapolated to inaccessible tissues, such as brain, heart, lung or pancreas. In these comparisons one needs to differentiate between intra-individual differences and inter-tissue correlations that are related to true epigenetic variants. Moreover, cross-tissue comparison studies should be undertaken in all stages of life. In future we might get answers why some regulatory variants are more critical than others for trait variability or disease risk. Understanding the epigenome, including the pathways of non-coding RNAs, would have a great potential in the prevention or treatment of diseases, such as cancer.

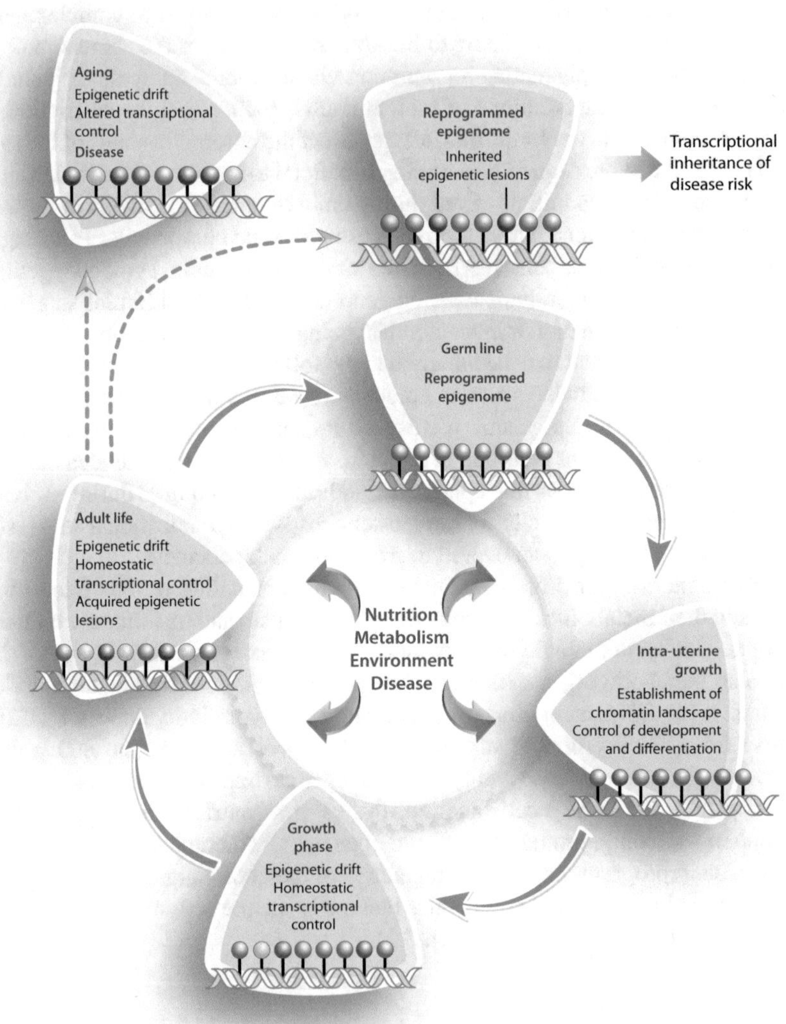

Fig. 5.8 Epigenetic drift and transgenerational inheritance of risk for metabolic disease. During embryogenesis epigenetic marks, such as DNA methylation and post-translational histone modifications are established, in order to maintain cell lineage commitment. After birth, this chromatin landscape stays dynamic throughout life and responds to nutritional, metabolic, environmental and pathological signals. This epigenetic drift is part of homeostatic adaptations and keeps the individual at good health. However, when an adverse epigenetic drift compromises the capacity of metabolic organs to adequately respond to nutritional or inflammatory challenges, susceptibility to metabolic disease increases. Some of these acquired epigenetic marks can be inherited to subsequent generations when they escape epigenetic reprogramming during gametogenesis

Individualized nutrition, tailored as optimal epigenetic diet, may be considered as future "medicine" for the maintaince of wellbeing and the therapy of age- and life-style-related diseases.

Key Concepts

- Epigenetics is the study of functionally relevant chromatin modifications that do not involve a change in the genome sequence.
- The change in DNA methylation during development starts with demethylation during cell devisions of the fertilized egg, followed by *de novo* methylation after implantation.
- The dynamic competition between nucleosomes and transcription factors for critical binding sites within genomic DNA is influenced by a large set of chromatin modifying enzymes that covalently modify histone proteins.
- A wide spectrum of plant-derived metabolites, such as genistein, resveratrol, curcumin and polyphenols from green tea, coffee and cocoa, are able to modulate *in vitro* the activity of transcription factors and chromatin modifiers.
- Numerous connections between products of intermediary metabolism and chromatin modifying enzymes are known. Since the cellular concentrations of several of these metabolites represent the metabolic status of the cell, the activities of the chromatin modifiers reflect the intermediary metabolism.
- A derivative of the B vitamin folate, tetrahydrofolate, serves as a methyl-group donor for feeding of the cyclic one-carbon pathway that is essential for DNA and histone methylation. The dependence of the pathway on folate and other micronutrients is another example of the direct connection between nutrition and epigenetics.
- The A^{vy}/a *Asip* mouse is an ideal *in vivo* model to investigate the link between environmental stimuli, such as nutrition, and epigenetic states of the genome.
- Exposure to sub-optimal nutrition during fetal development leads to changes in the germ-cell DNA methylome of male offspring.
- Different rodent models indicate that epigenetic memory can be passed from one generation to another by inheriting the same indexing of chromatin marks.
- Thrifty hypothesis: Fetal malnutrition leads to impaired fetal growth and low birth weight that favors a thrifty phenotype being epigenetically programmed to use nutritional energy efficiently.
- Developmental origins of health and disease hypothesis: The early-life development is critically sensitive to inadequate nutrition and other environmental factors leading to permanent changes in metabolism that can alter susceptibility to complex diseases.
- Nutrigenomic approaches will open up new therapeutic strategies, such as a possible reprogramming of the epigenome of metabolic organs through personalized diet including natural compounds that modulate the activity of chromatin modifiers and transcription factors.

Additional Reading

Carlberg C, Molnár F (2014) Mechanisms of gene regulation. Springer, Dordrecht. ISBN 978-94-007-7904-4

Gut P, Verdin E (2013) The nexus of chromatin regulation and intermediary metabolism. Nature 502:489–498

Heard E, Martienssen RA (2014) Transgenerational epigenetic inheritance: myths and mechanisms. Cell 157:95–109

Jirtle RL, Skinner MK (2007) Environmental epigenomics and disease susceptibility. Nat Rev Genet 8:253–262

Kaelin WG Jr, McKnight SL (2013) Influence of metabolism on epigenetics and disease. Cell 153:56–69

Mill J, Heijmans BT (2013) From promises to practical strategies in epigenetic epidemiology. Nat Rev Genet 14:585–594

Vanden Berghe W (2012) Epigenetic impact of dietary polyphenols in cancer chemoprevention: lifelong remodeling of our epigenomes. Pharmacol Res 65:565–576

Chapter 6
Nutritional Signaling and Aging

Abstract Under conditions of calorie restriction, i.e. at reduced food intake, the lifespan of model organisms, such as yeast, worms or flies, is increased. Similarly, when the activity of nutrient-sensing pathways is reduced by the knockdown of one or several of their key genes, these species live longer. Even in rodents and rhesus monkeys decreased nutrient-sensing pathway activity or calorie restriction protects them against T2D, CVD and cancer. The molecular basis of these processes are sensing of glucose and amino acids via the insulin/IGF and the TOR pathways, respectively, and the integration of the nutritional and energetic status of cells and tissues via HDACs of the sirtuin family and AMPK. Since these signal transduction pathways are evolutionary conserved, also humans may be protected against age-related pathologies. Humans may slow down their aging process, when their nutrient signaling pathways are modulated by moderate intake of a diet that was personalized for them.

In this chapter, we will get insight into the evolutionary conservation of nutrition-sensing pathways and their relation to the aging process. We will realize that higher organisms, such as mammals, use more complex regulatory circuits for sensing food that involve the CNS via the growth hormone endocrine axis. However, a detailed view on signal transduction related to calorie restriction will show us that even for humans very similar regulatory principles apply. We will discuss this insight and its potential application in preventing of age-related diseases and promoting healthy aging in humans.

Keywords Aging • Model organisms • Nutrient sensing • Insulin/IGF signaling • TOR-S6K signaling • Growth hormone endocrine axis • Calorie restriction • Sirtuins • Glucose metabolism • Lipid metabolism • NAD^+ • AMPK • Cellular energy status • AMP

C. Carlberg et al., *Nutrigenomics*, DOI 10.1007/978-3-319-30415-1_6

6.1 Aging and Conserved Nutrient-Sensing Pathways

Aging is a complex molecular process that affects almost all species. It is represented by the accumulation of molecular, cellular and organ damage, leading to loss of function and increased risk to disease and early death. Nutrient-sensing pathways are fundamental to the aging process. Abundance of food activates nutrient-sensing pathways that stimulate a diverse set of physiological processes, including reproduction, but this is compromised by a limited lifespan. In contrast, in times of starvation or when the nutrient-sensing pathways are genetically interrupted, reproduction is delayed and the lifespan increased. This suggests that the availability of food determines the speed of aging. Understanding the molecular basis of aging is a central topic of nutrigenomics. However, aging research in humans takes time, and for ethical reasons many types of experiments are not possible. Therefore, most of the principles of aging were first understood via the use of model organisms, such as *Saccharomyces cerevisiae* (yeast), *Caenorhabditis elegans* (roundworm), *Drosophila melanogaster* (fruit fly) and *Mus musculus* (mouse) that have a far shorter lifespan than humans (Box 6.1).

Box 6.1 Model Organisms

A model organism is a non-human species that is studied *in vivo*, in order to understand biological processes, such as aging, that due length of lifespan, costs or ethical reasons cannot be studied in humans (Fig. 6.1). The evolutionary conservation of biological pathways allows the transfer of at least some of the results and insights obtained with the model organism to humans. In the unicellular species yeast, not only the survival of a population of non-dividing cells (chronological lifespan) can be studied, but also the number of daughter cells generated by a single mother cell (replicative lifespan). The roundworm *C. elegans* is a simple multi-cellular species formed only by some 1000 cells but already allows studies of different cell types and organs, such as the nervous or digestive systems. It was the first organism, in which lifespan extending mutations were found. The fruit fly *D. melanogaster* confirms the evolutionary conservation of biological pathways. It contains more different tissues than *C. elegans* and allows the examination of sex differences. Finally, the mouse is the most established model organism in biomedical research and as a mammalian species it is much closer to humans (of course, primates are closest, but for ethical and generation length reasons, research with them is limited).

(continued)

Box 6.1 (continued)

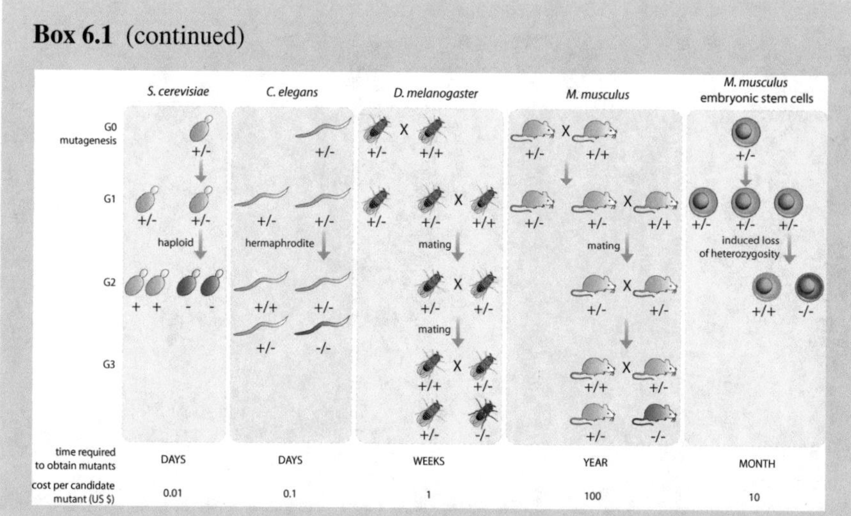

Fig. 6.1 Model organisms in aging research. Short generation length and the possibility to study genetic mutants with variant lifespans make the model organisms yeast, roundworm, fruit fly and mouse attractive for aging research

Multi-cellular organisms have developed a nutrient-sensing system allowing communication between different parts of the body. The intracellular signal transduction pathway of the peptide hormone IGF1 is the same as that of insulin, and informs cells about the presence of glucose (Sect. 9.1). The insulin/IGF pathway is evolutionary very conserved and, depending on the respective species, starts with one or several specific receptor tyrosine kinase-type membrane receptors (Fig. 6.2). Via cytosolic adaptor proteins, such as insulin receptor substrate (IRS), and kinases, such as phosphoinositide 3-kinase (PI3K) and Akt murine thymoma viral oncogene homolog (AKT), the ligand stimulation of the receptors results in the inactivation of one or several members of the forkhead box (FOX) transcription factor family that control the expression of genes involved in a wide range of physiological processes, such as cellular stress response, anti-microbial activity and detoxification of xenobiotics and free radicals (Sects. 9.2 and 9.3). At least in lower organisms, such as *C. elegans* and *D. melanogaster*, the knockout of any gene/protein that leads to the specific interruption of this signal transduction pathway causes lifespan extension.

Parallel to the glucose-sensing system there is an amino acid-sensing pathway that is also evolutionary highly conserved (Fig. 6.2). Central to this signal transduction pathway are the proteins TOR and S6 kinase (SK6). TOR is a highly conserved cell-growth modulator and serine/threonine kinase, which in mammalians exists in two different complexes, mammalian target of rapamycin complex 1 and 2 (mTORC1 and mTORC2), of which only mTORC1 is sensitive to amino acids (Sect. 3.1). The TOR-S6K pathway demonstrates cross-talk with the insulin/IGF

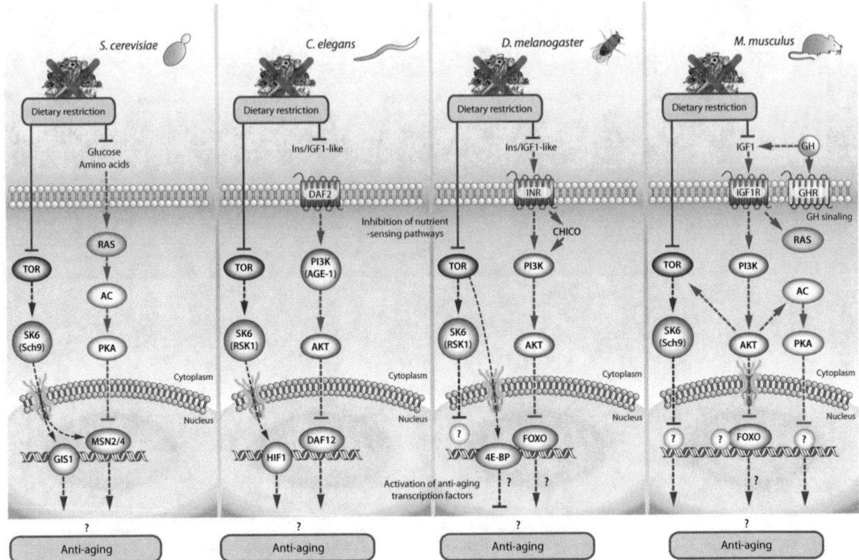

Fig. 6.2 Nutrient signaling pathways involved in longevity are conserved in various species. The activity of various signal transduction pathways is reduced by calorie restriction either directly (in yeast) or indirectly (in worm, fruit fly and mice) through the reduced levels of growth factors, such as IGF1. In all four species TOR and S6K activation promote aging (reduce lifespan). Moreover, in yeast and mammals activation the adenylate cyclase (AC)-protein kinase A (PKA) pathway accelerates aging. In addition, aging is promoted by the insulin/IGF1 signaling pathway directly or indirectly via its upstream factors, such as GH. Transcription factors, such as GIS1, MSN2/4, DAF16 and FOXO1, are inactivated by either the AC-PKA, the IGF1-AKT or the TOR-S6K pathway. They protect from aging in all the major model organisms

pathway and the inhibition of its activity also can increase lifespan, at least in lower organisms. In mice, mutations in genes for growth hormone (GH) and the insulin/IGF signaling pathway can increase lifespan by up to 50 %. Mice that are deficient in GH in their muscle cells and fibroblasts, express more anti-oxidant enzymes and have increased stress resistance than normal mice. In contrast, in liver, kidney, muscle and heart the administration of GH decreases the anti-oxidant defense of mice. Moreover, the inhibition of the TOR pathway increases the lifespan of mice and at the same time reduces the incidence of age-related pathologies, such as bone, immune and motor dysfunctions and insulin resistance.

Taken together, the study of conserved nutrient-sensing pathways suggests that the genetic alterations create a physiological state in the investigated model organisms that resembles to periods of food shortage. Interestingly, a reduction in food intake without malnutrition, referred to as calorie restriction (Sect. 6.3), is also able to extend the lifespan of diverse species spanning from yeast to rhesus monkeys.

6.2 Neuroendocrine Aging Regulation in Humans and Other Mammals

Aging starts for humans after the peak of their maximal physical ability in the age of 20–25 years, i.e. already at this age begins a slow progressive decline of the physiological capabilities of the different tissues and cell types forming the human body. From an age of 45 years onwards, there is a significant increase in the onset of aging-related complex diseases, such as cancer, T2D, CVD and Alzheimer's disease. The average generation length of modern humans for many thousand years was approximately 20 years and just increased in the recent past. Assuming some additional 25 years for raising all offspring, evolutionary adaption had only approximately 45 years to select humans for sufficient fitness and health. This means that any harm occurring to humans above this age, such as developing diabetes or cardiovascular problems, can not be corrected via evolutionary adaption principles, such as an increased or decreased number of vital offspring (Sect. 2.2). Nevertheless, the maximal lifespan of humans extends to approximately 120 years. The extra time of 75 years, however realized only in a few subjects, can be considered as a margin of safety, in order to guarantee that the vast majority of humans have enough time to fulfill the primary evolutionary sense of life, i.e. to reproduce and to assure the survival of their children until these start reproduction themselves.

Studies of the mouse model have indicated that in mammals the sensing of nutrition involves additional control circuits, including actions of the CNS and its associated glands. For example, the somatotrophic axis comprises GH (Sect. 6.1) that is secreted by the anterior pituitary and its secondary mediator IGF1 that is produced primarily by the liver. Interestingly, GH receptor-deficient primates develop seldom diabetes or cancer but due to developmental defects and the increased mortality at younger ages they do not seem to have an increased life expectancy. However, genetic variations that reduce the functions of GH, IGF1 receptor (IGF1R), insulin receptor (IR) or their downstream effectors, such as AKT, TOR and FOXO1, have been linked also to longevity in humans (Fig. 6.3). This reflects a defensive response of minimized cell growth and metabolism in case of cellular damage or food shortage, which enables an organism with a constitutively decreased insulin/IGF signaling to survive longer. For the same reason physiologically or pathologically aged mammals decrease their insulin/IGF signaling, i.e. during normal aging the levels of GH and IGF1 decline. In addition to insulin/IGF signaling sensing glucose, also in mammals high amino acid concentrations are sensed by TOR and low-energy states by sirtuins via high NAD^+ levels (Sect. 6.4) and by AMPK via high AMP levels (Sect. 6.5) (Fig. 6.3). During aging, TOR activity increases in mouse hypothalamic neurons and contributes to age-related obesity, which can be reversed by infusion of the TOR inhibitor rapamycin to the hypothalamus. However, TOR inhibition creates side effects, such as impaired wound healing and insulin resistance, i.e. this pathway is not suited for pharmacological intervention in humans. In contrast, sirtuins and AMPK are counteracting to insulin/IGF and TOR signaling, i.e. their activity represents low food availability and catabolism instead of nutrient

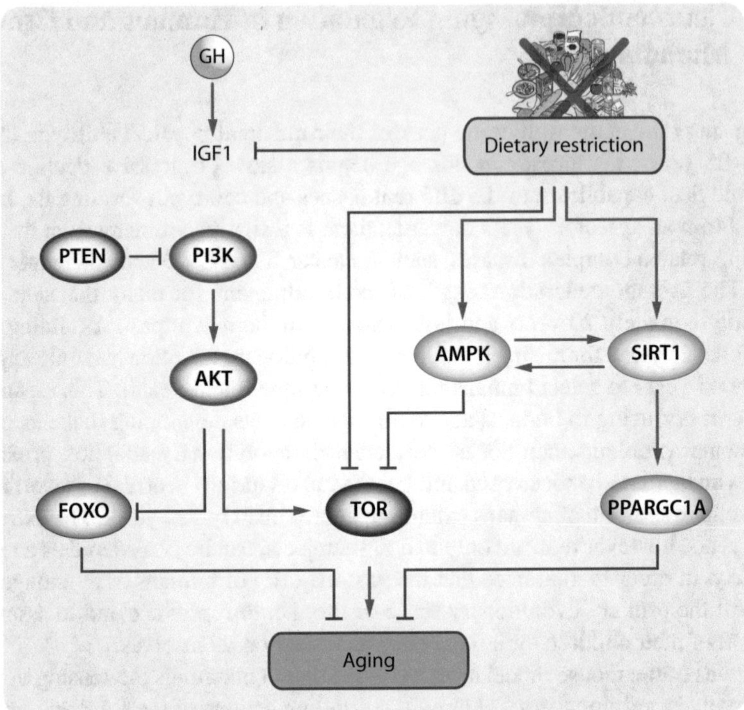

Fig. 6.3 Nutrient-sensing in mammals. Overview of the endocrine somatotroph axis that involves GH and the insulin/IGF1 signal transduction pathway. Proteins that favor aging are shown in orange and those with anti-aging properties in green

abundance and anabolism. Interestingly, one of the activities of AMPK is the direct inhibition of TOR. The sirtuin SIRT1 and the kinase AMPK are connected in a positive feedback loop concerning sensing low-energy states of cells. Furthermore, SIRT1 controls the activity of the co-activator protein PPARGC1A, which in turn manages central metabolic pathways counteracting aging, such as mitochondriogenesis, enhanced anti-oxidant defenses and improved fatty acid oxidation (Fig. 6.3) (Sect. 6.4). Taken together, anabolic signaling accelerates aging also in mammals, while decreased nutrient signaling extends the lifespan.

6.3 Calorie Restriction from Yeast to Mammals

For thermodynamic reasons life does not function without energy supply, i.e. there would be no life without high-energy nutrients, such as fatty acids or glucose. When lowering food intake of model organisms, such as yeast, worms or flies, their lifespan rises to a maximum but then declines rapidly. This indicates that there is a species-specific optimal percentage of dietary reduction and an amplitude of

	Life-span increase [fold]		Beneficial health effects	
	Dietary restriction	Mutations/ drugs	Dietary restriction	Mutations/ drugs
S. cerevisiae	3	10 with starvation	Extended reproductive period	Extended reproductive period, decreased DNA damage mutations
C. elegans	2-3	10	Resistance in misexpressed toxic proteins	Extended motility, resistance in mis-expressed toxic proteins and germ-line cancer
D. melanogaster	2	1.6-1.7	None reported	Resistance to bacterial infection, extended ability to fly
M. musculus	1.3-1.5	1.3-1.5 100% in combi-nantion with DR	Protection against cancer, diabetes, atherosclerosis, cardiomyopathy, autoimmune, kidney and respiratory diseases, reduced neurodegeneration	Reduced tumor incidence; protection against age-dependent cognitive decline, cardiomyopathy, fatty liver and renal lesions, extended insulin sensitivity
M. mulatta	trend noted	not tested	Prevention of obesity, protection against diabetes, cancer and cardiovascular diseases	not tested
H. sapiens	not tested	not tested GHR-deficient subjects reach old age	Prevention of obesity, diabetes, hypertension reduced risk factors for cancer and cardiovascular diseases	Possible reduction in cancer and diabetes

Fig. 6.4 Calorie restriction. In a number of model organisms experiments of calorie restriction have been performed, where nutrient-sensing pathways have been modulated genetically or chemically. However, there is a wide range of results and the long-term effects in humans are not yet known

response in terms of life extension. In general, calorie restriction inactivates one or several nutrient signaling pathways, such as the insulin/IGF1 or the TOR pathway (Fig. 6.4). During these periods of food scarcity, the organisms enter a standby mode, in which cell division and reproduction are minimized or even stopped, in order to save energy for maintenance systems putting survival in preference before reproduction. This means that most species have developed an anti-aging system, in order to overcome periods of starvation.

In humans and other mammals, extreme calorie restriction causes detrimental health effects, such as infertility and immune deficiencies. However, beneficial effects of moderate calorie restriction are obtained both by reducing carbohydrate intake as well as by reducing fat or protein consumption, i.e. also in mammals several nutrition sensing pathways respond to reduction in food intake. Calorie restriction can increase the lifespan of rodents by up to 60 %. In general, dietary-restricted rodents show many levels of metabolic, hormonal and structural adaptations when reducing body fat mass, higher insulin sensitivity as well as reduced inflammation and oxidative damage (Table 6.1). This is also observed in calorie-restricted monkeys. For example, in rhesus monkeys a 30 % calorie restriction over 20 years

Table 6.1 Effects of calorie restriction on mammalian tissues

Tissue	Effects on dietary restriction
Liver	Increase in gluoconeogenesis and glycolysis
	Decrease in glycolysis
Muscle	Increase in mitochondrial biogenesis and respiration
	Increase in β-oxidation of fatty acids
	Increase in protein turnover
Fat	Decrease in storage of triacylglycerols
	Decrease in secreted leptin
	Increase in secreted adiponectin
Pancreatic β cells	Decrease in secreted insulin
Brain	Decrease in pituitary secretion of growth hormone, thyroid hormone, gonadotropins
	Increase in adrenal secretion of corticoids
Whole organism	Increase in insulin sensitivity and decrease in blood glucose
	Increase in metabolism

reduced the incidence of cancer and CVD by 50 % compared with controls and there was even no case of diabetes or pre-diabetes. In analogy, also in humans, calorie restriction provides beneficial effects against obesity, insulin resistance, inflammation and oxidative stress. Moreover, also humans show adaptations in their hormonal circuits, such as increased levels of adiponectin and reduced concentrations of triiodothyronine, testosterone and insulin. Moreover, reduced concentrations of cholesterol and CRP are observed and also blood pressure decreases. Taken together, in rodents, monkeys and even in humans calorie restriction protects against age-related diseases, such as T2D, CVD and cancer.

Despite convincing evidence of health benefits from calorie restriction, in the general population this approach may not be socially and ethically accepted for reducing the risk of age-related diseases. An alternative may be repeated fasting and eating cycles, which, at least in rats, prolonged the lifespan by more than 80 %. Moreover, mice that got a high-fat diet with regular fasting breaks, were lean, had lower levels of circulating inflammatory markers and no fatty liver compared to mice that consumed an equivalent total number of calories *ad libitum*. The body of the anatomically modern human is genetically still adapted to a life as a hunter and gatherer and has not changed much since then (Chap. 2). Therefore, at stone age time the human feeding pattern was most likely also shifting between starvation and times of overeating after a successful hunt. However, the efficacy of most fasting strategies are probably limited, if they are not combined with diets that have health-associated benefits, such as the Mediterranean diet. Taken together, the switch between nutrient intake, usage and storage, i.e. between feeding and fasting, is a fine-tuned regulatory, evolutionarily conserved program that involves the nutrient-sensing insulin/IGF1 and TOR signaling pathways (Sects. 6.1 and 6.2) and the food restriction pathways involving sirtuins (Sect. 6.4) and AMPK (Sect. 6.5).

6.4 Properties and Functions of Sirtuins

The combination of nutritional epigenomics (Chap. 5) and the beneficial health effects of calorie restriction and feeding-fasting alternations (Sect. 6.3) suggests that proteins, such as the HDACs of the sirtuin family, may be the key molecules in this nutrigenomic process. The protein silent information regulator 2 (Sir2), a transcriptional silencer of mating-type loci, telomeres and ribosomal DNA in yeast, is the founding member of the "sirtuin" family of NAD^+-dependent HDACs. Sirtuins combine epigenetically important enzymatic activity with the sensing of the energy status of cells, i.e. the NAD^+/ NADH ratio. In humans, the family has seven members, SIRT1 to SIRT7 (Table 6.2). Sirtuins act in different cellular compartments. SIRT1, SIRT6 and SIRT7 are pre-dominantly localized in the nucleus but at least SIRT1 is also found in the cytosol. In contrast, SIRT2 is mainly cytosolic and enters the nucleus only during the G2 to M phase transition of the cell cycle. SIRT3, SIRT4 and SIRT5 are preferentially located in mitochondria. In the nucleus, the substrates of sirtuins are histones and transcription factors, but also in the cytoplasm and in mitochondria they remove acetyl-groups from post-translationally modified regulatory proteins. Interestingly, SIRT4 and SIRT6 can also function as ADP-ribosyltransferases (Table 6.2).

Table 6.2 Mammalian sirtuins

Sirtuins	Molecular mass [kDa]	Cellular localization	Activity	Key regulatory functions
SIRT1	81.7	Nucleus and cytosol	Deacetylase	Metabolism, inflammation
SIRT2	43.2	Cytosol	Deacetylase	Cell cycle and motility, myelination
SIRT3	43.6	Mitochondria	Deacetylase	Fatty acid oxidation, antioxidant defences
SIRT4	35.2	Mitochondria	ADP-ribosyl-transferase	Amino acid-stimulated insulin secretion, suppression of fatty acid oxidation
SIRT5	33.9	Mitochondria	Deacetylase, Demalonylase, Desuccinylase	Urea cycle
SIRT6	39.1	Nucleus	Deacetylase, ADP-ribosyl-transferase	Genome stability, metabolism
SIRT7	44.8	Nucleus	Deacetylase?	Ribosomal DNA transcription

The sub-cellular localization of sirtuins, their mode of action and their functions in different compartments are listed

SIRT1, the human homolog to Sir2, is one of the best-studied human signaling proteins. It regulates the metabolism of glucose and fat in response to energy level changes and therefore acts as a central control of the energy homeostasis network. Important non-histone target of SIRT1 are the tumor suppressor protein p53 and the co-activator protein PPARGC1A (Sect. 5.2) (Fig. 6.5). The deacetylation by SIRT1 activates PPARGC1A and induces downstream pathways that control gene expression in mitochondria. Furthermore, SIRT1 controls the acetylation of FOX transcription factors, such as FOXO1, that are in turn inhibited by the insulin/IGF pathway (Sect. 6.1). In addition, SIRT1 represses the transcription of uncoupling protein 2 (UCP2). This increases the yield of ATP production from glucose oxidation and the secretion of insulin from β cells. Importantly, SIRT1, SIRT3 and SIRT6 all suppress the transcription factor hypoxia-inducible factor 1α (HIF1α) leading to decreased glycolysis and increased oxidative metabolism. From the three mitochondrial sirtuins, SIRT3 is the major deacetylase of key regulatory proteins in metabolic homeostasis, such long-chain acyl-CoA dehydrogenase (LCAD, being involved in fatty acid oxidation), 3-hydroxy-3-methylglutaryl-CoA synthase 2 (HMGCS2, regulates the production of ketone bodies), IDH2 (TCA cycle), glutamate dehydrogenase (GDH, also TCA cycle) and superoxide dismutase 2 (SOD2, a key mitochondrial anti-oxidant enzyme) (Fig. 6.5). Interestingly, SIRT4 ADP-ribosylates GDH, thereby inhibiting its activity and blocking amino acid-induced insulin secretion, i.e. SIRT4 counteracts the activity of SIRT3.

LXR is a key transcription factor controlling the expression of genes involved in lipid synthesis, partly via the induction of the gene *SREBF1*. SIRT1 deacetylates and increases the activity of both LXR and SREBF1 targeting lipid metabolism, while SIRT6 serves as a negative regulator of triacylglycerol synthesis (Fig. 6.6). SIRT1 inhibits adipogenesis and enhances fat mobilization through lipolysis by suppressing *PPARG* expression via complex formation with the co-repressor proteins NCOR1 and NCOR2. The activities of SIRT3, SIRT4 and SIRT6 in the regulation of fatty acid oxidation in the liver resemble those of glucose metabolism (compare Figs. 6.5 and 6.6).

In a simplified view, the deacetylase activity of sirtuins counterbalances nutrient-driven protein acetylation. During fasting and/or exercise, the sirtuin co-factor NAD^+ levels rise in muscle, liver and WAT, while high-fat diet reduces the NAD^+/NADH ratio. Pharmacological targeting of sirtuins, in particular of SIRT1, is promising for the treatment of T2D. The search for further natural or synthetic sirtuin activators led to the identification of several compounds, of which resveratrol, a polyphenol found in red grapes and berries (Sect. 5.1), received most attention. Resveratrol improves mitochondrial activity and metabolic control in humans, i.e. it may also increase the health span of humans. The most prominent synthetic sirtuin activator is SRT1720, which protects against diet-induced obesity by improving mitochondrial function. However, both resveratrol and SRT1720 seem to not activate sirtuins directly, but at least resveratrol acts via AMPK (Sect. 6.5).

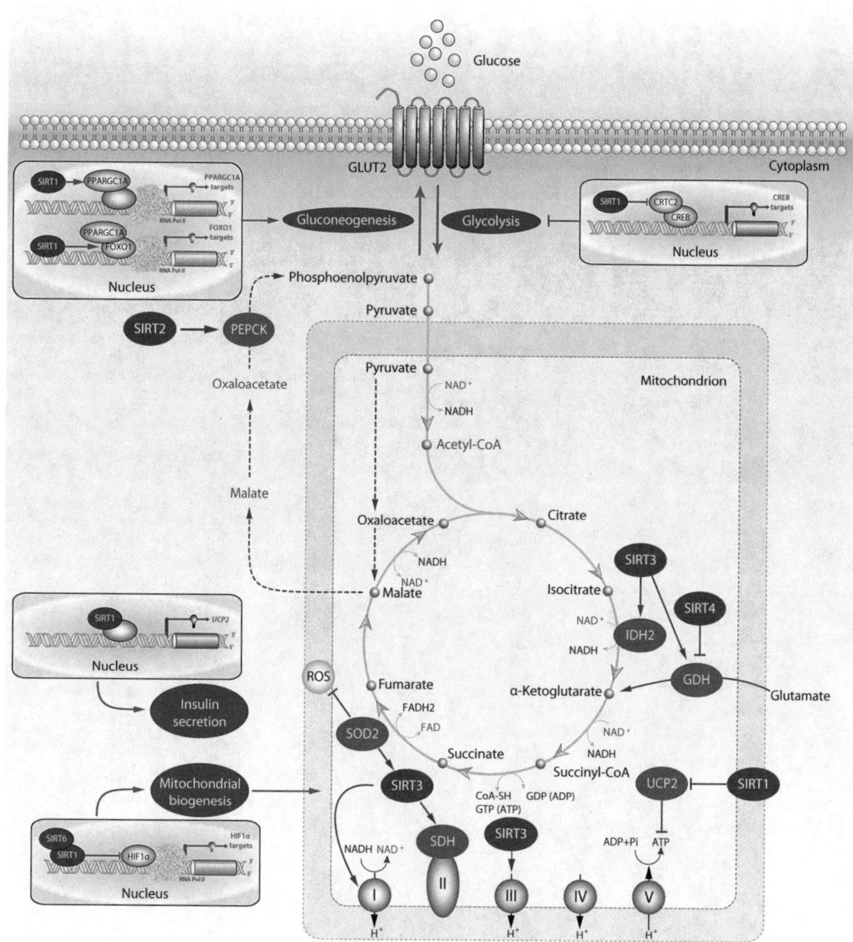

Fig. 6.5 Sirtuins and pathways involved in glucose metabolism. SIRT1 controls gluconeogenesis by deacetylating and activating the co-activator PPARGC1A and the transcription factor FOXO1. Furthermore, SIRT1 inhibits glycolysis, but activates lipid utilization and mitochondrial biogenesis. Moreover, it increases insulin secretion by suppressing uncoupling protein (UCP) 2 in the pancreas. Glucose taken up by the cell via GLUT is catabolized into pyruvate (glycolysis), which enters the TCA cycle in mitochondria, in order to generate energy. In cases of limited energy supplies, gluconeogenesis increases by the conversion of pyruvate to oxaloacetate and malate. Malate is shuttled into the cytoplasm and there converted back to oxaloacetate. The latter is used by PCK2 to produce phosphoenolpyruvate that is converted to glucose. SIRT1 also increases insulin secretion by repressing *UCP2* gene expression. SIRT2 activates gluconeogenesis by increasing the stability of PCK2. SIRT3 decreases ROS production by stimulating SOD2. Moreover, SIRT3 enhances oxidative phosphorylation by increasing the activities of complex I, complex II (via SDH), complex III and IDH2. By modulating the activity of GDH SIRT3 and SIRT4 both also regulate gluconeogenesis and insulin secretion through deacetylation and ADP-ribosylation, respectively

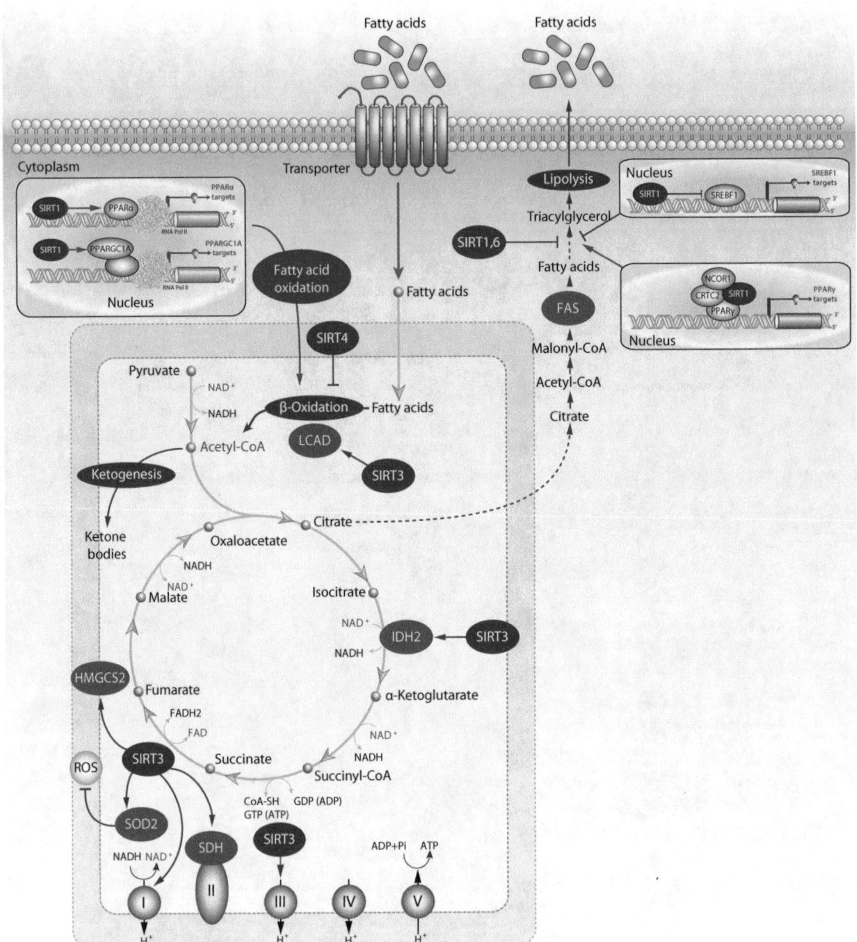

Fig. 6.6 Sirtuins control lipid metabolism. SIRT1 inhibits lipid synthesis by deacetylating SREBF1 through suppressing the activity of PPARγ. SIRT1 also promotes fatty acid oxidation by increasing *PPARA* and *PPARGC1A* gene expression. In mitochondria, fatty acids are oxidized, in order to produce ATP. Via the action of HMGCS2 this fatty acid breakdown can lead to the production of ketone bodies. Fatty acids can be synthesized in the cytosol from malonyl-CoA by FASN and then converted to triacylglycerols. In adipose tissue during periods of high-energy demand, triacylglycerols can be broken down by lipolysis into FFAs that are released into circulation. Moreover, SIRT1 decreases fatty acid storage by increasing lipolysis via inhibiting PPARγ and decreasing fatty acid synthesis via SREBF1. In addition, SIRT6 acts as a repressor of genes involved in fatty acid synthesis. SIRT3 increases β-oxidation and the formation of ketone bodies via LCAD and HMGCS2, decreases ROS production by stimulating SOD2 and enhances cellular respiration by increasing the activities of complex I, complex II, complex III and IDH2, respectively. SIRT4 dampens the expression of genes involved in fatty acid oxidation

6.5 Cellular Energy Status Sensing by AMPK

ATP is mainly generated by oxidative phosphorylation in the mitochondrial membrane. A high ATP/ADP ratio is essential to promote nearly all processes in living cells, since they all require energy and mostly are driven by the hydrolysis of ATP to ADP. Then, approximately every second ADP molecule is converted by the enzyme adenylate kinase to AMP. Therefore, falling cellular energy is associated with a decreased ATP/ADP ratio on one side and with increases in both ADP and AMP on the other side. Thus the energy status of a cell can be monitored either via the ATP/ADP or the ATP/AMP ratio. The ATP/AMP ratio is directly sensed by a few enzymes, such as glycogen phosphorylase and phosphofructokinase (both get activated) and fructose-1,6-bisphosphatase (which gets inhibited). However, the main sensor of the energy status of cells is AMPK. AMPK is a heterotrimeric complex of its catalytic α-subunit and the regulatory β- and γ-subunits (Fig. 6.7) and is activated

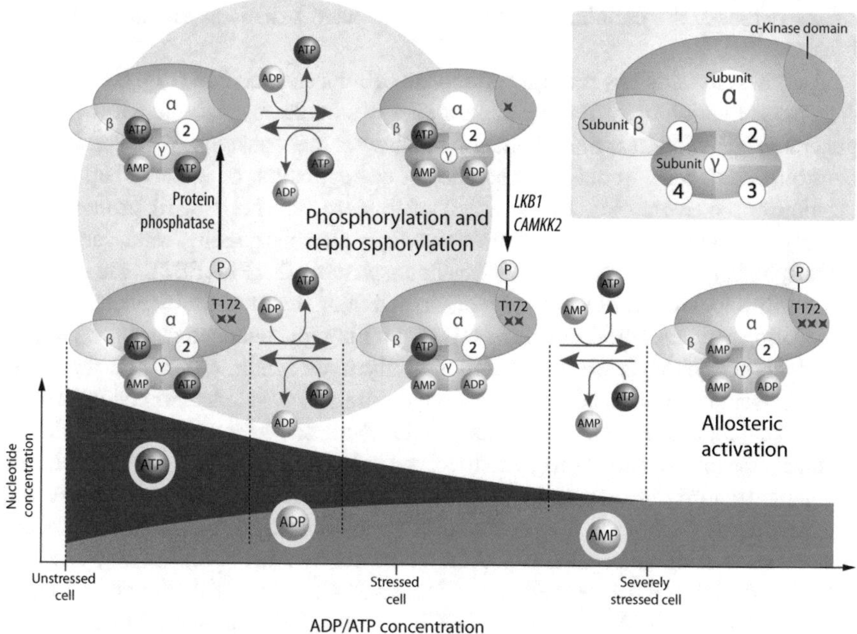

Fig. 6.7 Model of the mechanism for AMPK activation. Increasing cellular concentrations of AMP and ADP activate AMPK. In the basal state ATP binds to sites 1 and 3 of the γ-subunit of AMPK, while AMP always occupies site 4. When during moderate stress ATP is replaced by ADP or AMP at site 3 the phosphorylation of T172 is stimulated, which causes a 100-fold increase in the respective activity. During more severe stress ATP is replaced by AMP at site 1 and thus causes a further tenfold activation. When the cellular energy status returns to normal, AMP at site 1 and either ADP or AMP at site 3 are progressively replaced by ATP. This stimulates the dephosphorylation of T172 and a return to the basal state. Changes in the predicted concentrations of ATP, ADP and AMP are displayed. They go from an unstressed, fully charged cell to a cell undergoing a severe energy stress corresponding to a tenfold decrease in the ATP/ADP ratio. In contrast, the AMP concentration in a fully charged cell is very low, but when ADP/ATP increases its percentage change in concentration is significantly larger than those of ATP or ADP

by various types of metabolic stresses, drugs and xenobiotics involving increases in cellular AMP, ADP or Ca^{2+}. The major upstream kinases of AMPK are liver kinase B1 (LKB1) and Ca^{2+}/calmodulin-dependent protein kinase kinase 2 (CAMKK2). LKB1 provides a high basal level of AMPK phosphorylation that is modulated by the AMP binding, while the alternative activation pathway via CAMKK2 responds to increases in cellular Ca^{2+} but is independent of AMP or ADP level changes.

AMPK is activated by stress that inhibits the catabolic production of ATP, such as starvation for glucose or oxygen, as well as by stress that increases ATP consumption, such as muscle contraction. Furthermore, numerous synthetic compounds, such the anti-diabetic drugs metformin, phenformin and thiazolidinediones, as well as plant products, such as resveratrol (wine), epigallocatechin gallate (green tea), capsaicin (peppers) and curcumin (turmeric) activate AMPK. However, the activation by most of these compounds, including metformin and resveratrol, is indirect via the increase of cellular AMP and ADP through the inhibition of mitochondrial ATP synthase. Since these AMPK activators also extend the lifespan of *C. elegans*, they may act as mimetics of calorie restriction and/or exercise. Both processes decrease the cellular energy status and have beneficial effects on healthy lifespan.

AMPK activation has multiple effects on cellular metabolism (Fig. 6.8). In general, the activation of AMPK by lack of energy stimulates catabolic pathways that generate ATP, while it turns off anabolic pathways that consume ATP. The AMPK-controlled catabolic processes include i) up-regulation of glucose uptake by promoting the expression and function of glucose transporters, ii) promotion of glycolysis under anaerobic conditions by phosphorylating and activating 6-phosphofructo-2-kinase/fructose-2,6-biphosphatase 2 (PFKFB2), the enzyme responsible for the synthesis of glycolytic activator fructose 2,6-bisphosphate, iii) stimulating mitochondrial biogenesis, iv) inhibiting gluconeogenesis and v) reducing glycogen synthesis via the inhibition of glycogen synthase (GS). Furthermore, AMPK decreases fatty acid synthesis by inhibiting acetyl-CoA carboxylase (ACC) 1, the key regulatory enzyme in fatty acid synthesis. Moreover, AMPK down-regulates the expression of enzymes involved in fatty acid synthesis and inhibits the lipogenic transcription factor SREBF1. Furthermore, AMPK promotes uptake and β-oxidation of fatty acids in mitochondria via inhibition of ACC2. Mitochondrial biogenesis is another important process activated by AMPK and is mediated via SIRT1 and PPARGC1A as well as via increased clearance of dysfunctional mitochondria via UNC-51-like kinase 1 (ULK1). Finally, this generates increased capacity for the oxidative catabolism of both fatty acids and glucose. AMPK also conserves ATP by inactivating anabolic pathways, such as the biosynthesis of lipids, carbohydrates, proteins and ribosomal RNA. AMPK can down-regulate the expression of the proteins involved in these pathways. Moreover, AMPK also influences whole-body metabolism and energy balance via mediating effects of hormones and other agents acting on neurons of the primary appetite control center, the arcuate nucleus of the hypothalamus, that regulates intake of food and energy expenditure actions (Sect. 8.6).

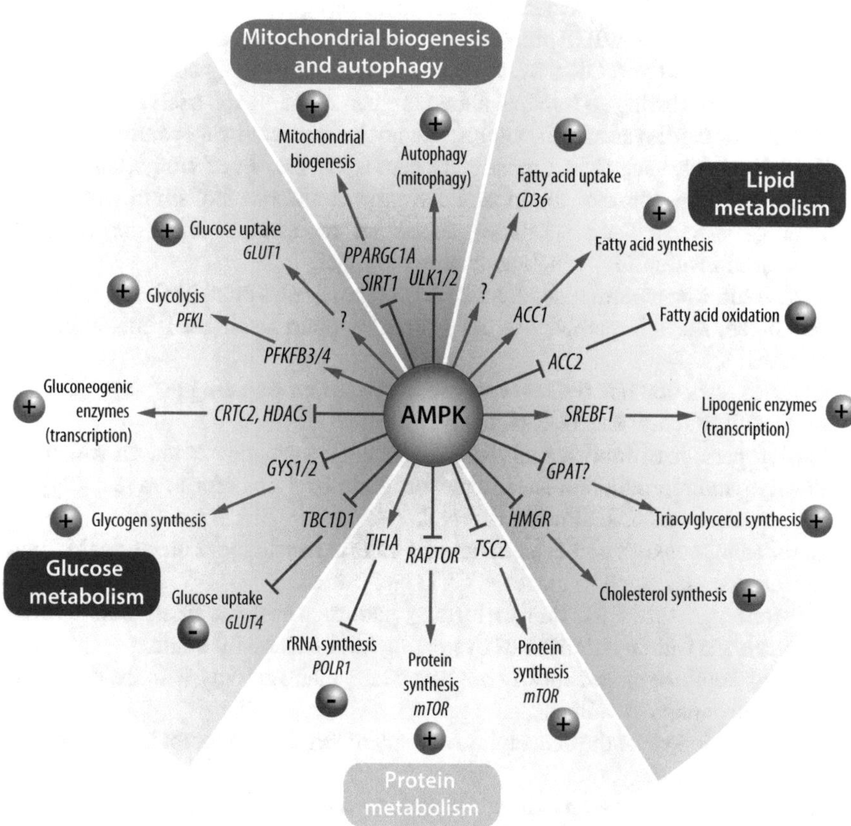

Fig. 6.8 Consequences of AMPK activation. This scheme displays the metabolic effects of AMPK activation. Proteins shown with question marks may not be directly phosphorylated by AMPK. CRTC2, CREB-regulated transcription co-activator 2; GPAT, glycerol phosphate acyl transferase; RAPTOR, regulatory associated protein of mTOR; TBC1D, TBC1 domain; TIFIA, transcription initiation factor IA; TSC2, tuberous sclerosis 2

Future View

In future, aging may be delayed and age-related diseases may be prevented by reprogramming gene expression of major nutrition signaling pathways. However, more important than extending the lifespan will be a diet-triggered prolongation of the healthy lifespan. Although diet shows convincing heath benefits, it may not be the ideal approach for humans. In future, alternating fasting-feeding regimes and natural or synthetic sirtuin activators may be used for the treatment and prevention of human diseases.

Key Concepts

- In case of food availability, activated nutrient-sensing pathways stimulate a diverse set of activities, including reproduction, but this is compromised by a limited lifespan.

- The availability of food determines the speed of aging.
- Parallel to the insulin/IGF glucose sensing system there is an amino acid sensing pathway formed by TOR/S6K. Both are evolutionary highly conserved.
- Any harm occurring to humans above the age of 45 years, such as developing diabetes or cardiovascular problems, can not be corrected via evolutionary adaption principles, such as an increased or decreased number of vital offspring.
- A defensive response of minimized cell growth and metabolism in case of cellular damage or food shortage enables an organism with a constitutively decreased insulin/IGF signaling to survive longer.
- TOR inhibition creates side effects, such as impaired wound healing and insulin resistance, i.e. this pathway is not suited for pharmacological intervention in humans.
- Calorie restriction inactivates one or several nutrient signaling pathways, such as the insulin/IGF1 or the TOR pathway.
- During periods of food scarcity, organisms enter a standby mode, in which cell division and reproduction are stopped or minimized, in order to save energy for maintenance systems allowing survival.
- In rodents, monkeys and even in humans calorie restriction protects against age-related diseases, such as diabetes, CVD and cancer.
- At stone age times the human feeding pattern was most likely also shifting between starvation and times of overeating after successful hunts.
- Sirtuins combine epigenetically important enzymatic activity with the sensing of the energy status of cells.
- In a simplified view, the deacetylase activity of sirtuins counterbalances nutrient-driven protein acetylation.
- Resveratrol improves mitochondrial activity and metabolic control in humans, i.e. it may increase the human health span.
- The main sensor of the energy status of cells is AMPK.
- AMPK is activated by stresses that inhibit the catabolic production of ATP, such as starvation for glucose or oxygen, as well as by stresses that increase ATP consumption, such as muscle contraction.
- Activation of AMPK by energetic stress stimulates catabolic pathways that generate ATP, while it turns of anabolic pathways that consume ATP.

Additional Reading

Fontana L, Partridge L, Longo VD (2010) Extending healthy lifespan – from yeast to humans. Science 328:321–326

Hardie DG, Ross FA, Hawley SA (2012) AMPK: a nutrient and energy sensor that maintains energy homeostasis. Nat Rev Mol Cell Biol 13:251–262

Houtkooper RH, Pirinen E, Auwerx J (2012) Sirtuins as regulators of metabolism and healthspan. Nat Rev Mol Cell Biol 13:225–238

Lopez-Otin C, Blasco MA, Partridge L, Serrano M, Kroemer G (2013) The hallmarks of aging. Cell 153:1194–1217

Chapter 7
Chronic Inflammation and Metabolic Stress

Abstract Macrophages are associated with various tissues and either derive from monocytes circulating in the blood or from self-renewing embryonal cell populations. They show a large variety of stimulus- and tissue-specific functions, of which the extremes are pro-inflammatory M1-type and anti-inflammatory M2-type macrophages. M1 macrophages are key cells in the initiation of the acute inflammatory response, while M2 macrophages are resolving inflammation and coordinate tissue repair. However, tissue inflammation is not only caused by bacterial infection or tissue injury but may also derive from changes in the concentration of nutrients and metabolites. In this case, the immune system cannot cope the primary stimulus, so that chronic inflammation develops. This metabolic stress, in contrast to infectious or traumatic stress, is often caused by lipid overload in the blood and in adipose tissue. This again is a hallmark of age-related metabolic diseases, such as obesity, insulin resistance and atherosclerosis. For example, hypercholesterolemia (Sect. 11.3) causes stress to macrophages and their associated cells. Moreover, perturbations of the homeostasis of nutrient metabolism dys-regulate functions of the liver.

In this chapter, we will present monocytes and macrophages as the key players in acute and chronic inflammation. We will provide molecular and cellular details of examples of metabolic stress, such as disturbance of reverse cholesterol transport and ER stress. In this context, we will discuss macrophages as important therapeutic targets.

Keywords Monocytes • M1- and M2-type macrophages • Dendritic cells • Cytokines • Acute inflammation • Chronic inflammation • Cholesterol crystals • Inflammasome • Reverse cholesterol transport • LDL • HDL • Metabolic stress • ER stress • LXR • PPAR

7.1 The Central Role of Monocytes and Macrophages

Monocytes constitute up to 10 % of circulating leukocytes and are produced in the bone marrow from common myeloid progenitors of granulocytes and monocytes, referred to as colony-forming units (M-CFUs) (Fig. 7.1). The differentiation of myeloid progenitors is principally regulated by cytokines and orchestrated via the cooperative action of transcription factors, such as SPI1, CEBPs and AP-1. After

C. Carlberg et al., *Nutrigenomics*, DOI 10.1007/978-3-319-30415-1_7

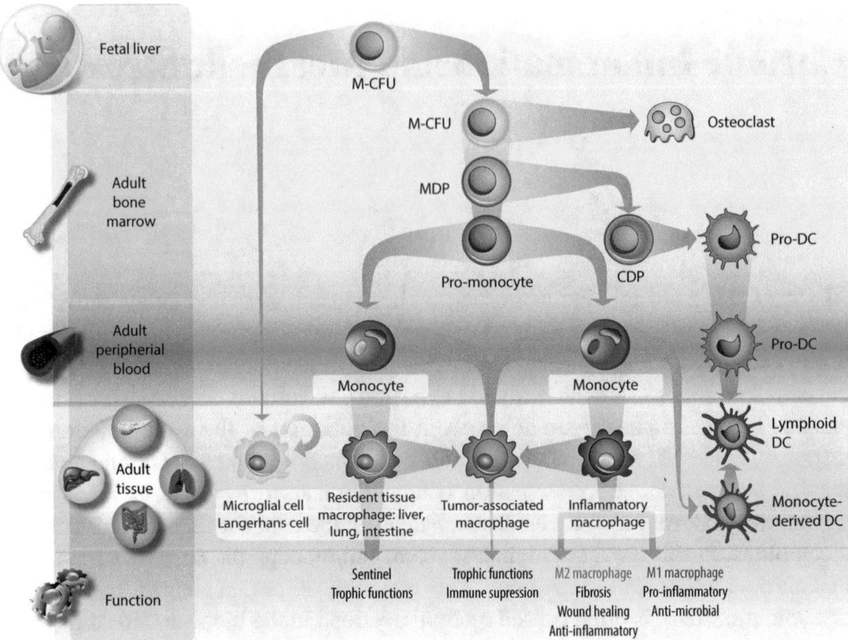

Fig. 7.1 Differentiation of monocytes. M-CFUs in the bone marrow are the precursors to macrophages and dendritic cell progenitors (MDPs). In the bone marrow, MDPs differentiate to common dendritic cell progenitors (CDPs) or to pro-monocyte precursors. Langerhans cells in the skin, microglial cells in the brain and a number of other tissue-resident macrophages initially develop during embryogenesis from M-CFUs in the yolk sac or fetal liver. The remaining tissue macrophages get polarized, depending on the inflammatory milieu, into M1 or M2 macrophages (Fig. 7.2). Monocytes are also the precursors to monocyte-derived dendritic cells

differentiation, monocytes are released into the blood stream and migrate within 1–3 days through the endothelium of blood vessels into tissues. This extravasation is stimulated by the cytokine- and chemokine-induced expression of adhesion molecules, such as selectins, on the surface of endothelial cells. In tissues, monocytes differentiate into macrophages or dendritic cells, i.e. into the central cellular components of the innate immune system (Box 2.2). Thus, the primary role of monocytes is to refill the pool of tissue-resident macrophages and dendritic cells in steady state and in response to inflammation.

Tissue-resident macrophages have a wide variety of homeostatic and immune surveillance functions ranging from the clearance of cellular debris, the response to infection and the resolution of inflammation. A substantial proportion of these macrophages is self-renewing and derives during embryogenesis from the yolk sac and the fetal liver. Populations of tissue-resident macrophages are found in most tissues of the human body, such as Kupffer cells in the liver, microglia in the CNS, osteoclasts in the bone and alveolar macrophages in the lung. Transcriptional profiling indicates that the same tissue often contains several phenotypically distinct macrophage subsets that have tissue-specific functions. Nevertheless, the unifying role of all tissue-resident macrophages is that they are in the frontline of the immune defense. In response to their activation by pathogens or metabolites, macrophages secrete a number of signaling molecules, such as cytokines, chemokines and growth factors, which affect the migration and activity of other immune cells. This response is called acute inflammation, when it is caused by infection or injury, and is often associated with erythema, hyperthermia, swelling and pain. However, it resolves within a few days to weeks. In contrast, low-grade chronic inflammation, such as in obesity (Sect. 8.6), does not cause heat or pain, but it can last over months and years, when the origin of the stimulus cannot be resolved (Sect. 7.2).

For simplicity, macrophages are often grouped into two classes, M1 and M2 macrophages, which rather represent two extremes of a continuum of functional profiles (Fig. 7.2). Pro-inflammatory molecules, such as interferon gamma (IFNγ), TNF and TLR activators, induce M1-type macrophages that in turn secrete further pro-inflammatory molecules, in order to sustain the inflammatory reaction. This classical pathway of IFNγ-dependent macrophages activation provokes the adaptive immune system to respond through the proliferation of type 1 T_H cells (T_H1) (Box 7.1).

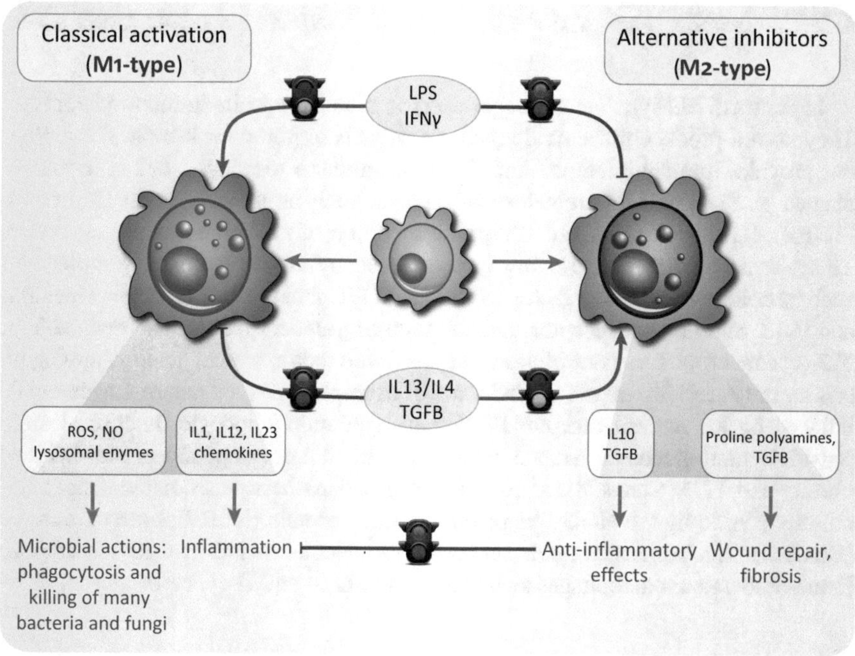

Fig. 7.2 Classical and alternative macrophage activation. Different stimuli activate monocytes/macrophages to develop into functionally distinct populations. Classically activated macrophages (M1-type) are induced by cytokines and microbial products, particularly IFNγ, and are microbicidal as well as involved in potentially harmful inflammation. Alternatively activated macrophages (M2-type) are induced by IL4 and IL13 produced by T_H2 cells and are important in tissue/wound repair and fibrosis

Box 7.1 Subsets of T Lymphocytes.
T cells constitute up to 30% of all circulating leukocytes and are a major component of the adaptive immune system (Box 2.2). They occur in a number of important subtypes, such as T-helper- (T_H) and cytotoxic T cells. T_H cells are characterized by the expression of the CD4 glycoprotein on their surface and support other cells of the immune system, such as maturation of B cells into antibody producing plasma cells and memory B cells as well as the activation of cytotoxic T cells and macrophages. Antigen-presenting cells, such as dendritic cells, activate T_H cells via the presentation of microbe-derived peptides on major histocompatibility complex (MHC) II receptors. Depending on their cytokine expression profile T_H cells are subdivided into T_H1 (IFNγ and IL2), T_H2 (IL4, IL5 and IL13) and T_H17 (IL17) cells. Moreover, T_H cells can differentiate into T_{REG} cells that are involved in immune tolerance and inhibit overboarding immune responses. Cytotoxic T cells (also called T killer cells) are characterized by CD8 expression and can destroy virus-infected cells and tumor cells. They get activated via the presentation of peptides (derived from intra-cellular proteins) on MHC I receptors on the surface of all nucleated cells.

In contrast, M2-type macrophages exert an almost opposite immuno-phenotype. They do not produce nitric oxide (NO) or radicals required for killing of microbes but provoke immunotolerance and T_H2-type immune responses. M2-type macrophages produce anti-inflammatory molecules, such as tumor growth factor beta (TGFB), IL10 or IL1RN, and inhibit the secretion of pro-inflammatory cytokines. This alternate macrophage pathway is induced by anti-inflammatory molecules, such as colony stimulating factor 2 (CSF2), TGFB and the T_H2-type cytokines IL4 and IL13, and nuclear receptor ligands, such as glucocorticoids. The main role of M2-type macrophages is resolution, such as tissue repair, wound healing, angiogenesis and extracellular matrix deposition. For example, M2-type macrophages can be induced by the nuclear receptor PPARγ and maintain adipocyte function, insulin sensitivity and glucose tolerance, which prevents the development of diet-induced obesity and T2D (Chaps. 8 and 10). However, when obesity-associated danger signals are sensed by the NOD-like receptor (NLR) protein (NLRP) 3 inflammasome (Box 7.2), this protein complex serves as a molecular switch that let the adipose tissue-associated macrophages switch from an M2 to an M1-type phenotype.

7.2 Acute and Chronic Inflammation

Acute inflammatory responses to insults, such as injury and infection, are critical for organism's health and recovery. Infection or tissue damage is initially sensed by PRRs, such as TLRs, NLRs, retinoic acid-inducible gene 1 (RIG1)-like helicase receptors (RLRs), lectins and scavenger receptors, that bind to PAMPs and DAMPs

Box 7.2 The Inflammasome

Inflammasomes belong to the key components of the intracellular surveillance system. The inflammasome is a large protein complex composed of NLRPs, the adaptor protein apoptosis-associated speck (ASC) and the pro-inflammatory caspase (CASP) 1 and CASP5 that is formed in response to the rise of the levels of a number of pathogen-associated molecular patterns (PAMPs) and damage-associated molecular patterns (DAMPs) (Fig. 7.3). Inflammasome activation requires a priming signal from PPRs, such as TLR4, and a second signal involving potassium ion efflux, lysosomal damage or ROS generation. Cholesterol crystals in lysosomes can provide this second signal, either as a result of phagocytosis of extracellular cholesterol crystals or via uptake of modified LDLs and free cholesterol released from LDLs. Inflammasome activation leads to the secretion of the pro-inflammatory cytokines IL1B and IL18. NLRP3 is the most prominent NRLP.

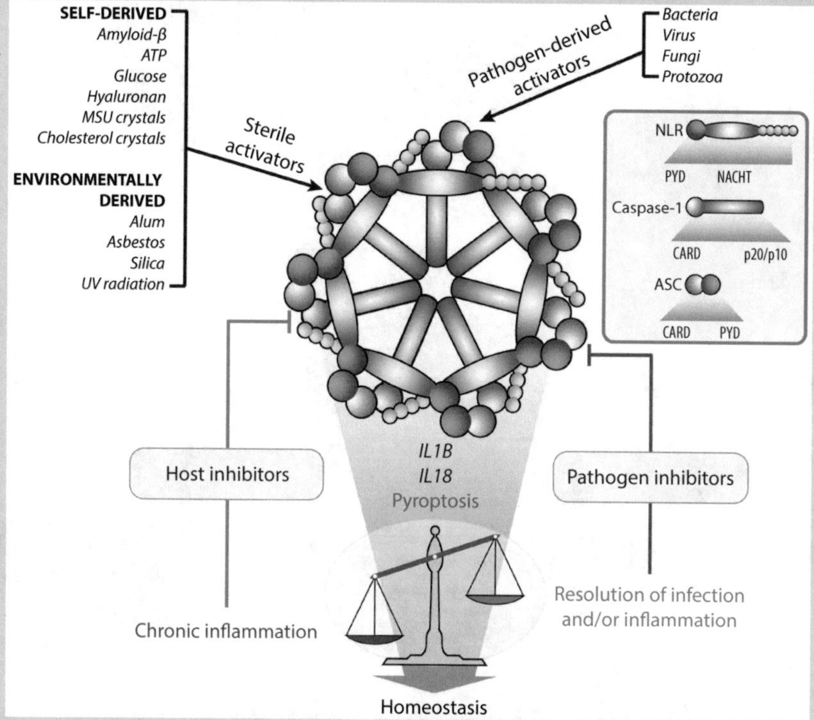

Fig. 7.3 Central role of the inflammasome. During acute or chronic inflammation, inflammasomes are directly or indirectly activated by a wide array of DAMPs. The initial event leads to the activation of CASP1, the release of IL1B and IL18 as well as sometimes to pyroptosis, which is a form of apoptosis associated with anti-microbial responses during inflammation. The release of IL1B and IL18 induces the recruitment of effector cell populations of the immune response and tissue repair, i.e. the activation of inflammasomes results in the resolution of infection or inflammation and contributes to homeostatic processes. However, constant activation of the inflammasome can lead to chronic inflammatory diseases. Pathogen-derived inhibitors can block inflammasome activation and thus the resolution of infection, while host-derived inflammasome inhibitors will prevent chronic inflammation

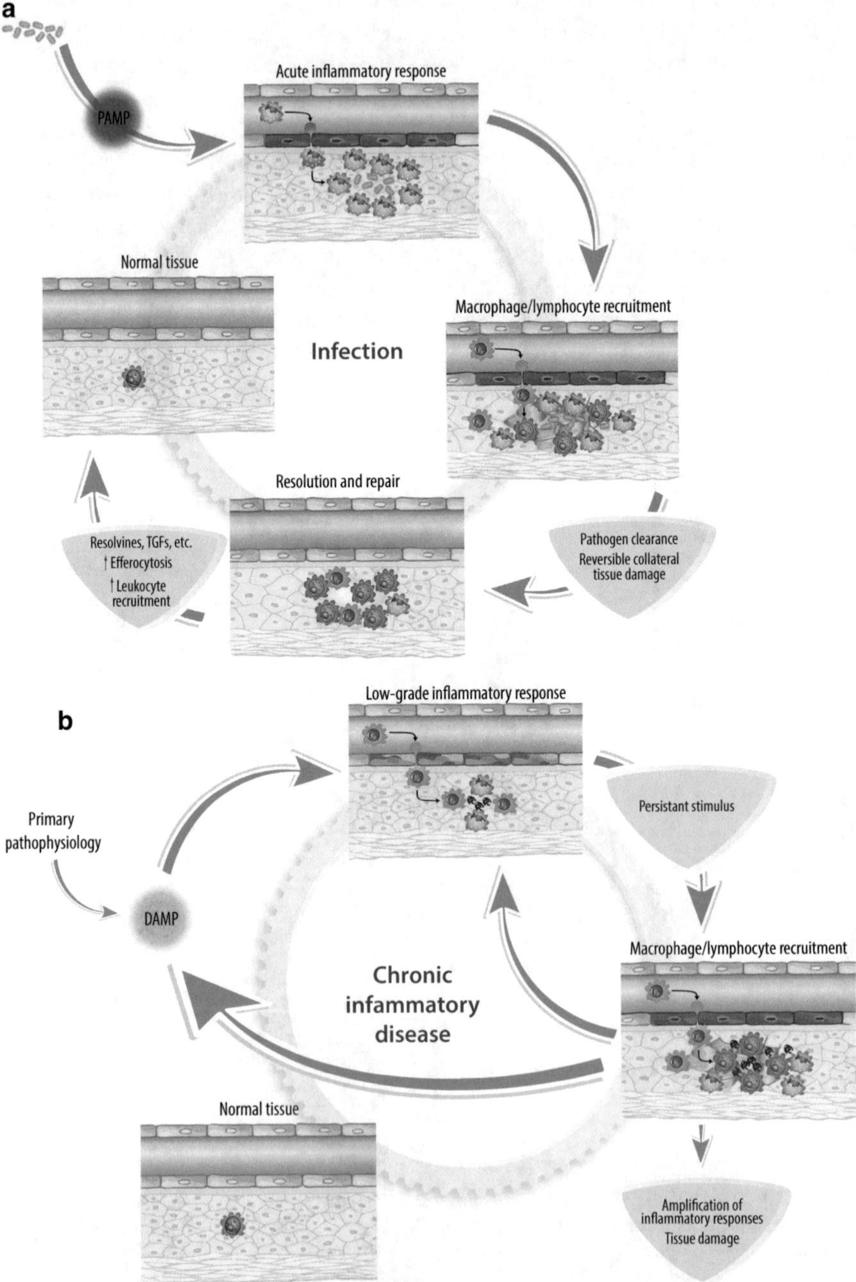

Fig. 7.4 Acute and chronic inflammation. (**a**). Acute inflammatory response to infection is initiated by the presentation of PAMPs to PRRs. Eradication of the pathogen eliminates the stimulus but may cause some collateral but reversible tissue damage. The resolution/repair phase leads then to

(Fig. 7.4a). A tissue-specific use of these receptors implicates that distinct macrophage subsets are involved in the activation of the immune response to microbes. However, basically in all cases PAMP- or DAMP-activated PRRs start a signal transduction cascade that ends in the activation of inflammation-associated transcription factors, such as NF-κB and AP-1. Major targets of these transcription factors are genes encoding for pro-inflammatory cytokines, such as *TNF* and *IL1B*, anti-microbial factors, such as inducible nitric oxide synthase 2 (*NOS2*), and cell-recruiting chemokines, such as *CCL2* and *CCL5*. Another important component of the DAMP sensing system is the inflammasome (Box 7.2) that enhances IL1B production and secretion. After initial recognition of the microbial challenge, resident macrophages stimulate the influx of cells from the blood, such as neutrophils and monocytes as a source of inflammatory macrophages (Sect. 7.1).

Most inflammatory lesions are initially dominated by monocyte-derived macrophages. The changed expression profile of these macrophages activates further cells of the innate immune system as well as the adaptive immune system. This implies that the early inflammatory response comprises a number of redundant components and is further amplified in cytokine-mediated feed-forward loops. Under resting conditions most tissues contain only a few resident macrophages, while at acute inflammation the number of immune cells drastically increases and the characteristics of these cells changes, respectively. For example, adipose tissue of lean persons contains deactivated macrophages, eosinophils and T_{REG} cells, while within the initial phase of the acute inflammation there is rapid invasion of neutrophils that is followed by the recruitment of monocyte-derived macrophages, T cells and stromal cells (Sect. 8.6).

The combined action of innate and adaptive immune cells eradicates the infectious microbes but also results in collateral tissue damage, such as cytotoxicity due to ROS production and the degradation of extracellular matrix via proteases. After the clearance of pathogens and the removal of apoptotic neutrophils by phagocytes, there is recruitment or phenotypic switching of macrophages into the M2-type, i.e. into a pro-resolving phenotype. This results in an overall repair and normalization of the tissue architecture and function, including the re-establishment of the vascularization. Most of these processes are activated by metabolites, such as prostaglandin E2 (PGE2) and ω-3 fatty acids. These metabolites function as signaling molecules binding to membrane proteins, such as GPR120, or directly activate nuclear receptors and initiate intracellular pathways leading to an anti-inflammatory response. This also involves anti-inflammatory cytokines, such as IL10 and TGFB1, and small lipid mediators, such as lipoxins, resolvins, protectins and maresins that

Fig. 7.4 (continued) the restoration of normal tissue homeostasis. (**b**). Chronic inflammation is caused by non-immune pathophysiological processes that trigger an initial sterile inflammatory response involving DAMPs and is amplified by cytokines and chemokines. However, this response does not eliminate the initial stimulus, so that non-resolving inflammation persists and results in continuous tissue damage

are produced by the enzymes arachidonate 5- and 15-lipoxygenase (ALOX5 and ALOX15) from arachidonic acid and ω-3 fatty acids.

In humans, the basal inflammatory response increases with age and leads to low-grade chronic inflammation that is maladaptive and further promotes the aging process. This may be due to (i) the accumulation of senescent cells that secrete pro-inflammatory cytokines, (ii) the increased likelihood that a failure of the immune system does not effectively clear pathogens and dysfunctional host cells, (iii) the overactive transcription factor NF-κB or (iv) an defective autophagy response. In all these cases, not microbes but the excess of endogenous molecules, such as lipoproteins, SFAs or protein aggregates initiate the inflammatory response (Fig. 7.4b). These alterations result in an enhanced activation of the NLRP3 inflammasome (Box 7.2) and other pro-inflammatory pathways, finally leading to the increased production of IL1B, TNF and interferons. Since metabolic dys-regulation accompanies aging, most common age-related diseases, such as T2D and CVD, are associated with chronic inflammation. The inflammatory response itself may amplify the production of disease-specific DAMPs. This results in positive-feedback loops that accelerate the underlying disease process (Sect. 7.3). For example, inflammation stimulates the formation of oxidized phospholipids that can serve as DAMPs in atherosclerosis (Sect. 11.2). Therefore, an inhibition of chronic inflammation can reduce the rate of disease progression to a point of substantial clinical benefit, although it does not alter the underlying pathogenic process.

7.3 Reverse Cholesterol Transport and Inflammation

The protective role of macrophages against antigenic substances, such as pathogenic microbes, enables them to recognize also lipid molecules, such as cholesterol crystals, when these accumulate within tissues. This process links macrophages to metabolism and assigns the cells with critical roles in atherosclerosis and lipid storage diseases. There are a number of interactions between inflammation and metabolism, such as that pathogens and inflammation disturb metabolic processes as well as that metabolic diseases lead to abnormal immune reactions. A central example is the process of reverse cholesterol transport. The transporter proteins ABCA1 and ABCG1 in the membrane of cholesterol-loaded macrophages interact with the main HDL protein APOA1 and mediate the efflux of cholesterol from the arterial wall to HDLs in the circulation (Fig. 7.5). The enzyme lecithin-cholesterol acyltransferase (LCAT) esterifies free cholesterol in HDLs. In the liver some of the free cholesterol or cholesteryl esters are selectively taken up via the transporter protein SCARB1 without degrading the HDLs. The liver either recycles the cholesterol as a component of secreted lipoproteins or excretes them as bile acids via the action of the transporters ABCG5 and ABCG8. In the plasma, the transporter protein CETP transfers cholesteryl esters from HDLs to VLDLs and LDLs in exchange for triacylglycerols. The enzymes LPL and hepatic lipase (LIPC) hydrolyze the

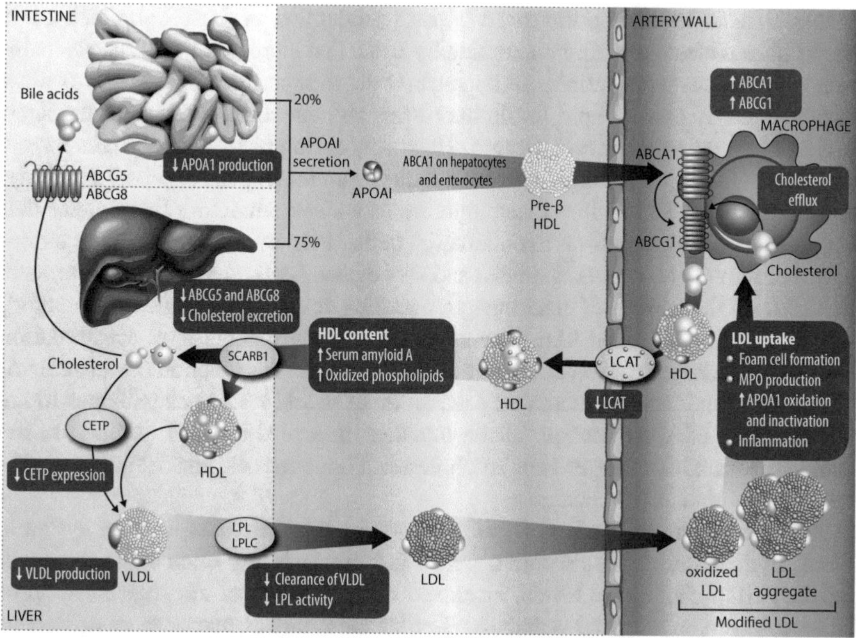

Fig. 7.5 Reverse cholesterol transport and the effect of inflammation. After secretion of APOA1 from the liver and the intestines, this protein interacts with ABCA1 on hepatocytes and enterocytes and is assembled into pre-βHDLs. In macrophages, these lipoproteins are loaded via the action of ABCA1 and ABCG1 with cholesterol and phospholipids and form HDLs. LCAT catalyzes the esterification of free cholesterol in HDLs. In the liver, SCARB1 removes some of the free cholesterol or cholesteryl esters from HDLs. In the liver, cholesterol is then either recycled into VLDLs via the action of CETP or is transformed into bile acids that are excreted by the transporters ABCG5 and ABCG8. VLDLs undergo a lipolytic cascade mediated by the enzymes LPL and hepatic lipase and are transformed into cholesterol-rich LDLs. A small proportion of the LDLs end up in the arterial wall, where they are modified by oxidation or aggregation and are taken up by macrophages. This leads to the formation of macrophage foam cells (Sect. 11.2), the production of myeloperoxidase (MPO) and inflammation. Red boxes indicated how inflammation negatively affects this reverse cholesterol transport

triacylglycerols of VLDLs, so that primarily cholesteryl esters remain and the lipoproteins transform into LDLs.

Increased LDL-cholesterol levels in the circulation are the principal drivers of atherosclerosis, since this causes cholesterol accumulation and inflammatory response in the artery wall (Fig. 7.5). Under healthy conditions, HDLs can oppose this process and are able to reduce inflammation by promoting the cholesterol efflux from foam cells. However, acute inflammation impairs reverse cholesterol transport at two central steps. On one hand it down-regulates the expression of *ABCA1* and *ABCG1* in macrophages, which reduces cholesterol efflux from macrophages, and on the other hand it decreases the hepatic expression of the genes *APOA1*, *CETP*, *ABCG5*, *ABCG8* and *CYP7A1*, which results in the reduced excretion of cholesterol. Moreover, acute inflammation causes the accumulation of triacylglycerols within VLDLs leading to increased hepatic production of these lipoproteins and their reduced clearance from circulation by LPL. The increased VLDL levels maintain high lipid levels in peripheral tissues, in order to suppress infection and to allow tissue repair. Increased lipid levels may even be beneficial, at least for the short period of an acute microbe infection. However, during acute sepsis HDLs have a decreased ability to mediate cholesterol efflux from macrophages (Fig. 7.5). Under these conditions, HDLs do not act anymore as anti-inflammatory lipoproteins that suppress monocyte adhesion to endothelial cells but in contrast support the monocyte recruitment. Moreover, oxidized and aggregated LDLs can activate PPRs, such as TLRs, on the surface of macrophages and thus directly trigger the inflammatory response. These modified LDLs are taken up and cause cholesterol accumulation within macrophages, which in turn amplifies the signal transduction downstream of TLRs. This leads via the increased production of cytokines and chemokines to the amplification of inflammation. Taken together, in a feed-forward mechanism the acute inflammation changes cellular cholesterol homeostasis, which further amplifies the inflammatory response.

Chronic infections, such as HIV1, and autoimmune diseases, such as systemic lupus erythematosus, rheumatoid arthritis and psoriasis, are often associated with reduced HDL-cholesterol levels, increased concentrations of atherogenic lipoproteins and accelerated atherosclerosis. Similar mechanisms may also contribute to other metabolic disorders. For example, on adipose tissue macrophages of obese persons TLRs and NLRs are activated by lipids, such as ceramides or SFAs. This can lead to chronic inflammation, insulin resistance and NAFLD (Sect. 9.4). During chronic inflammation, stimuli in atherosclerotic lesions, such as intracellular cholesterol crystals that activate the NLRP3 inflammasome (Fig. 7.4), activate the enzyme MPO that oxidizes APOA1 in HDLs and in this way further compromise the capability of the lipoproteins concerning reverse cholesterol transport. In individuals with CHD, the levels of oxidized APOA1 inversely relate to the ABCA1 efflux capacity and positively correlate with atherosclerotic disease (Sect. 11.2). Therefore, the inflammatory response of macrophages leads to local inactivation of APOA1 in the arterial wall and reduces cholesterol efflux from macrophages. In turn, this causes cholesterol accumulation in macrophages and further enhances the inflammatory responses. Moreover, HDL-cholesterol levels decrease and their composition changes, and through the oxidation of APOA1 they become dysfunctional. In addition, in progressive atherosclerotic plaques the interaction of LDLs with foam cells leads to an increased signaling of the receptors TLR2 and TLR4 and the cytosolic adaptor protein myeloid differentiation primary response protein 88 (MYD88) promoting cytokine and chemokine gene expression.

The inflammatory response in macrophages that is primarily mediated via TLR3 and TLR4, down-regulates the expression of the LXR target genes *ABCA1* and *ABCG1* (Sect. 3.4) through the induction of the transcription factor interferon-regulatory factor 3 (IRF3). Although this reduces cholesterol efflux, promotes cholesterol accumulation and enhances the inflammatory response, the increased levels of cholesterol result in higher oxysterol concentrations and respective LXR activation. In this way, increased cholesterol levels can stimulate their own efflux. Thus, the innate immune system can use changes in cholesterol metabolism, in order to amplify the inflammatory response, but also for restoring homeostasis.

7.4 M1 and M2 Macrophages in Adipose Tissue

Metabolic organs, such as the liver, pancreas and adipose tissue, are formed by parenchymal cells and stromal cells, including macrophages. In healthy state, these cell types work together, in order to maintain metabolic homeostasis. Also in disease these tissues try to interact, in order to adapt to altered conditions, such as increased nutritional needs of the affected organs. For example, during microbe infection, activated immune cells have an increased consumption of glucose, because they preferentially use the anaerobic but energetically inefficient glycolysis for ATP production. Therefore, these cells are supported best when the cytokines TNF, IL6 and IL1B are produced through the inflammatory response of macrophages and thus induce a timely restricted peripheral insulin resistance (Sect. 9.4), in order to decrease glucose storage. In fact, many diseases with active inflammatory responses, such as hepatitis C, HIV and rheumatoid arthritis, can lead to insulin resistance. Of note, the initiation and maintenance of immunity is energy consuming. For example, fever leads to a 7–13 % increase of energy consumption for every 1 °C of body temperature elevation, and sepsis can increase the human metabolic rate even by even 30–60 %.

Stromal cells, such as macrophages, support adipocytes in WAT in their main metabolic function, the long-term storage of lipids. The number and activation state of macrophage both reflect the metabolic health of WAT. In lean persons, only 10–15 % of the stromal cells are macrophages and most of them are of M2-type (Fig. 7.6). These M2-type macrophages secrete IL10 that potentiates insulin action in adipocytes, i.e. it maintains or even increases the insulin sensitivity of these cells. During the development of obesity, at least in mice WAT recruits monocytes that differentiate into M1-type macrophages and finally can comprise up to 60 % of all stromal cells in the tissue. These M1-type macrophages are a major driver of insulin resistance in WAT, but they are also involved in the remodeling of the enlarging adipocytes. Taken together, the two types of macrophages coordinate homeostatic adaptations of adipocytes in the lean and the obese state.

The T_H2-type cytokines IL4 and IL13 that are secreted primarily by eosinophils within WAT are essential to polarize macrophages into the M2-type (Fig. 7.6). The transcriptome profile of these macrophages is controlled by the nuclear receptors PPARδ and PPARγ and the Krüppel-like transcription factor (KLF) 4. Although adipocytes in lean persons can easily accommodate acute changes in food intake, a chronic overload with nutrients causes metabolic stress. As a result, the enlarging WAT releases chemokines, such as CCL2, CCL5 and CCL8, in order to recruit monocytes. These monocytes differentiate into macrophages that phagocytize dead adipocytes. The resulting lipid overload of the macrophages initiates, via the transcription factors IRF3 and NF-κB, the expression of pro-inflammatory cytokines,

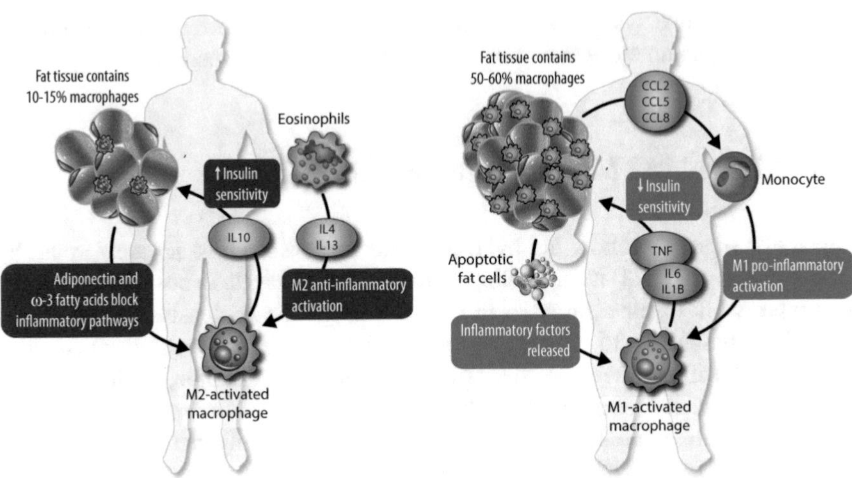

Fig. 7.6 M1 and M2 macrophages coordinate homeostatic adaptations of adipocytes. In lean healthy individuals, adipose tissue macrophages are primarily of M2-type. These are important for maintaining the insulin sensitivity of adipocytes, for example through the production of IL10. T_H2 cytokines, such as IL4 and IL13, which can be produced by a variety of cell types, such as eosinophils, are critical for the maintenance of the M2 phenotype in lean tissues. In contrast, during obesity, monocytes are recruited, which largely increase the number of macrophages. These M1-type macrophages are pro-inflammatory, secrete TNF, IL6 and IL1B and decrease insulin sensitivity while facilitating the storage of excess nutrients. In turn, expanding WAT releases chemokines, such as CCL2, CCL5 and CCL8, in order to recruit further inflammatory monocytes that intensify the process

such as TNF and IL6. Although the cytokine expression in obese tissues is significant, it is often rather modest and local in comparison to the acute inflammatory response after an infection or trauma.

Nevertheless, the long-term exposure of adipocytes with pro-inflammatory cytokines from M1-type macrophages can induce insulin resistance of WAT (Sect. 9.4). Like in acute microbe infection, this reduced insulin sensitivity initially tries to react to the increased levels of nutrients by limiting their storage. However, the strategy of inducing insulin resistance becomes maladaptive in case of a constant, long-term nutrient overload. Taken together, the hallmarks of obesity-induced inflammation, sometimes also referred as "metaflammation", are that it (i) is a nutrient-induced inflammatory response orchestrated by WAT-associated macrophages, (ii) changes the polarization of these macrophages from M2 to M1 phenotype, (iii) represents a moderate/low-grade and local expression of inflammatory cytokines and (iv) is chronic without apparent resolution.

In other metabolic organs, polarized macrophages play a comparable role. In the primary thermogenic organ, the BAT, resident macrophages differentiate into M2-type after exposure to cold temperatures. These M2 macrophages induce thermogenic genes in BAT and lipolysis of stored triacylglycerols in WAT via the secretion of the catecholamine noradrenaline. Kupffer cells, the resident macrophages of liver, enable the metabolic adaptations of hepatocytes during increased caloric intake. The M2 phenotype of Kupffer cells is induced via PPARδ and the T_H2-type cytokines IL4 and IL13. Under the condition of obesity the M2 macrophages regulate oxidation of fatty acids in the liver and support in this way hepatic lipid homeostasis. Similar to WAT, also in the pancreas high-fat diet induces the infiltration of M1-type macrophages. The increased intake of dietary lipids results in β cell dysfunction, which induces the expression of chemokines recruiting inflammatory macrophages to the islets. The secretion of IL1B and TNF by the infiltrating macrophages further augments β cell dysfunction (Sect. 9.5).

7.5 ER Stress Response

The ER is an organelle that has a vital role in maintaining cellular and whole body metabolic homeostasis. Although the ER has significant adaptive capacity to manage the periodic cycles associated with feeding, fasting and other metabolic demands of limited duration, it is less flexible to respond appropriately to chronic and escalating metabolic challenges. Therefore, ER dysfunction of cells of metabolic tissues, such as liver, pancreas, WAT and muscle, is a key contributor to metabolic disintegration and chronic inflammation (Sects. 8.6 and 9.5). For example, hepatocytes and adipocytes of obese humans show increased ER stress compared with lean controls. Moreover, multiple cardiovascular disease risk factors, such as inflammation, dyslipidemia, hyper-homocysteinemia and insulin resistance, can lead to the development of ER stress in atherosclerotic lesions.

The ER is the primary site of the 3-dimensional folding of all membrane proteins and secreted proteins that are produced in the cell and also performs their quality control. An accumulation of unfolded proteins in the ER as well as other challenges of ER functions, such as hypoxia (i.e. oxygen supply deprivation), infections, toxins, nutrient overload or energy deprivation, can trigger an adaptive, protective mechanism, the so-called unfolded protein response (Fig. 7.7a).

On the molecular level the ER stress response is primarily mediated by the proteins eukaryotic translation initiation factor 2-alpha kinase 3 (EIF2AK3), endoplasmic reticulum to nucleus signaling 1 (ERN1) and activating transcription factor 6 (ATF6). In the absence of a stress signal, these three transmembrane proteins are bound by the chaperone HSPA5 and are kept inactive. An increased protein load, in particular improperly folded proteins, activates EIF2AK3 and ERN1, i.e. they dissociate from HSPA5 and initiate signal transduction cascades. EIF2AK3 phosphorylates eukaryotic translation initiation factor 2A (EIF2A) and suppresses general protein translation. ERN1 interacts with TNF receptor-associated factor 2 (TRAF2) and activates the kinases inhibitor of kappa light polypeptide gene enhancer in B cells (IKBK) and mitogen-activated protein kinase 8 (MAPK8, also called JNK). This activates the transcription factors NF-κB and AP-1 and increases the expression of inflammatory cytokines. In metabolic tissues, such as WAT and

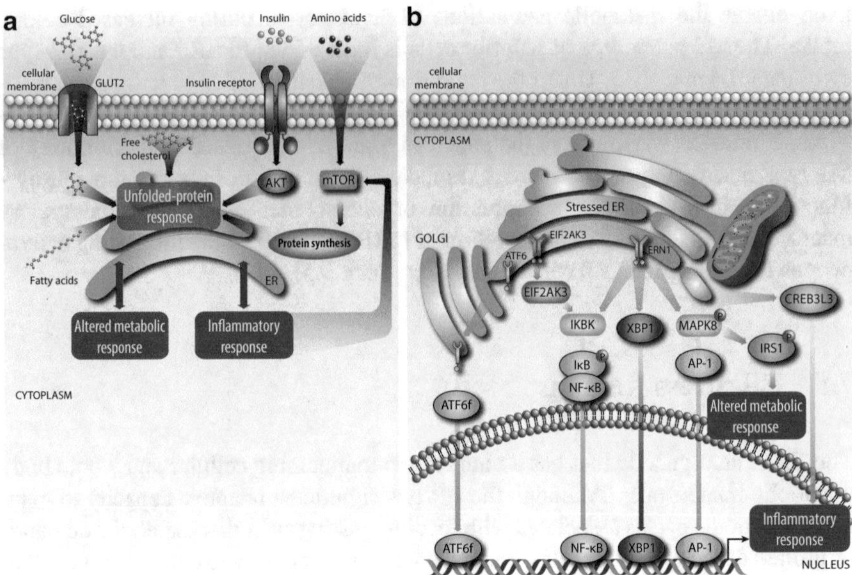

Fig. 7.7 The unfolded protein response and inflammation. (**a**) The ER responds to multiple nutrient-associated signals, such as those induced by fatty acids, glucose, free cholesterol, insulin and amino acids. ER stress due to nutrient overload induces the unfolded protein response, which activates inflammatory signaling pathways that result in altered metabolic and inflammatory responses. (**b**) The three ER transmembrane proteins EIF2AK3, ERN1 and ATF6 are the molecular mediators of the unfolded protein response. Via the phosphorylation of the kinases IKBK and MAPK8 (also called JNK) ERN1 activates the transcription factors NF-κB and AP-1 leading to an increase in the expression of inflammatory genes. This response is further enhanced by (i) translocation of ATF6 to the Golgi and processing there to an active transcription factor, (ii) up-regulation of the expression of the transcription factor XBP1 and (iii) the transcription factor CREB3L3. MAPK8 also phosphorylates IRS1 resulting in an altered metabolic response. A functional and molecular integration between the different organelles can mediate the spread of the stress

liver, the activation of MAPK8 also leads via serine phosphorylation of IRS1 to defective insulin actions, such as insulin resistance (Sect. 9.4).

Moreover, ERN1 has also endoribonuclease activity and cleaves, for example, the mRNA of the transcription factor X-box binding protein 1 (*XBP1*), which results in the translation of an activated form of XBP1 responsible for up-regulation of many chaperone genes. Furthermore, ATF6 translocates from the ER to the Golgi apparatus, where it is processed by proteases to an active transcription factor. Finally, ER stress also leads to the cleavage and activation of the transcription factor cAMP responsive element binding protein 3-like 3 (CREB3L3), which induces, particularly in the liver, the production of the acute-phase protein CRP. The goal of activating the three arms of the unfolded protein response is to restore ER homeostasis by (i) reducing general protein synthesis, (ii) facilitating protein degradation and (iii) increasing the protein folding capacity (Fig. 7.7b).

The lipophilic environment of the large ER membrane is provided with important functions in the metabolism of lipids, in particular of phospholipids and cholesterol. For example, cholesterol sensing is initiated at the ER membrane through SREBF1 (Sect. 3.1). This indicates a direct connection between lipid metabolism and the unfolded protein response, such as the control of ER phosphatidylcholine synthesis and ER membrane expansion by XBP1. Moreover, ER stress is linked to the production of inflammatory mediators, such as the enzyme prostaglandin-endoperoxide synthase 2 (PTGS2, also known as COX2) and ROS. This disturbs lipid metabolism and glucose homeostasis leading to abnormal insulin action, promotes hyperglycemia through insulin resistance, stimulates hepatic glucose production and suppresses glucose disposal. When the unfolding protein response cannot reconstitute proper ER function or when the metabolic stress continues, apoptotic pathways are initiated, i.e. the affected cells are dying. This happens, for example, to macrophage foam cells during atherosclerosis (Sect. 11.2).

Taken together, nutrient and inflammatory responses are integrated in metabolic homeostasis, but dysfunction of the ER affects this integration and results in chronic metabolic disease. For example, the reciprocal regulation between ER stress and insulin signaling pathways leads to a vicious cycle explaining the inter-dependence of insulin resistance (Sect. 9.4) and atherosclerosis (Sect. 11.2).

Future View

In order to widen the benefit-to-risk window of anti-inflammatory therapy in chronic diseases in future, it has to be found out, whether there is a chance to remove inflammatory stimuli, such as LDLs in atherosclerosis or nutrient excess in obesity. Moreover, the close connection between the molecular pathways of metabolic and inflammatory response imposes the questions, (i) whether it is possible in future to separate these two pathways, (ii) will there be non-inflammatory nutrients, i.e. can specific nutrients trigger the inflammation associated with metabolic stress, (iii) can alternative interventions preserve or recover metabolic health despite the presence of an inflammatory response and (iv) are there even beneficial effects from inflammation associated with metabolic diseases? Very selective activators of the nuclear receptors LXR and PPAR may be able to shift M1 macrophages into the anti-inflammatory M2 phenotype and to induce in this way a dominant program of resolution.

Future therapies of atherosclerosis-associated diseases may address the key suppressor of inflammation, IL10. In order to avoid systemic side effects of a long-term treatment with this cytokine, it may be delivered by nanoparticles to atherosclerotic lesions. In order to monitor the anti-inflammatory or pro-resolving actions of drugs, it will be important to develop methods that monitor with minimal invasion the different macrophages subsets.

Key Concepts

- The primary role of monocytes is to refill the pool of tissue-resident macrophages and dendritic cells in steady state and in response to inflammation.
- Acute inflammation is caused by an infection or an injury and resolves within a few days to weeks, while chronic inflammation lasts over months or years and origins from a stimulus that cannot be resolved.

- The main role of M1-type macrophages is to secrete pro-inflammatory molecules, in order to sustain the inflammatory reaction, while the function of M2-type macrophages is resolution, such as tissue repair, wound healing, angiogenesis and extracellular matrix deposition.
- During acute or chronic inflammation, inflammasomes are directly or indirectly activated by a wide array of DAMPs leading to the secretion of the pro-inflammatory cytokines IL1B and IL18.
- Under resting conditions, most tissues contain only a few resident macrophages, while at acute inflammation the number of immune cells drastically increases and the characteristics of these cells changes.
- Chronic inflammation is caused by non-immune pathophysiologic processes that trigger an initial inflammatory response involving DAMPs and being amplified by cytokines and chemokines.
- Since aging is accompanied by metabolic dys-regulation, most common age-related diseases, such T2D and CVD, are associated with chronic inflammation.
- There are a number of interactions between inflammation and metabolism, such as that pathogens and inflammation disturb metabolic processes as well as that metabolic diseases lead to abnormal immune reactions.
- In a feed-forward mechanism, the acute inflammation changes cellular cholesterol homeostasis, which further amplifies the inflammatory response.
- Increased cholesterol levels can stimulate their own efflux from macrophages and the eventual suppression of TLR-mediated inflammatory responses, i.e. the innate immune system uses changes in cholesterol metabolism to amplify the inflammatory response and then to restore homeostasis.
- The number and activation state of macrophages reflect the metabolic health of WAT.
- The two types of macrophages coordinate homeostatic adaptations of adipocytes in both the lean and the obese state.
- The hallmarks of metaflammation are that it (i) is a nutrient-induced inflammatory response orchestrated by WAT-associated macrophages, (ii) changes the polarization of these macrophages, (iii) represents a moderate/low-grade and local expression of inflammatory cytokines and iv) is chronic without apparent resolution.
- ER dysfunction of cells of metabolic tissues, such as liver, macrophages, WAT and muscle, is a key contributor to metabolic disintegration and chronic inflammation.
- The unfolded protein response is triggered by an accumulation of unfolded proteins in the ER as well as other challenges of ER functions, such as hypoxia, infections, toxins, nutrient overload or energy deprivation.
- The goal of activating the three arms of the unfolded protein response is to restore ER homeostasis by (i) reducing general protein synthesis, (ii) facilitating protein degradation and (iii) increasing the protein folding capacity.
- When the unfolding protein response cannot reconstitute proper ER function or when the metabolic stress continues, apoptotic pathways are initiated.

- In metabolic homeostasis, nutrient and inflammatory responses are integrated, but dysfunction of the ER affects this integration and results in chronic metabolic disease.

Additional Reading

Fu S, Watkins SM, Hotamisligil GS (2012) The role of endoplasmic reticulum in hepatic lipid homeostasis and stress signaling. Cell Metab 15:623–634

Lawrence T, Natoli G (2011) Transcriptional regulation of macrophage polarization: enabling diversity with identity. Nat Rev Immunol 11:750–761

Tabas I, Glass CK (2013) Anti-inflammatory therapy in chronic disease: challenges and opportunities. Science 339:166–172

Tall AR, Yvan-Charvet L (2015) Cholesterol, inflammation and innate immunity. Nat Rev Immunol 15:104–116

Wynn TA, Chawla A, Pollard JW (2013) Macrophage biology in development, homeostasis and disease. Nature 496:445–455

Part III
Links to Diseases

Chapter 8
Obesity

Abstract Obesity is the consequence of excess WAT accumulation that increases the risk of metabolic diseases (Chap. 12). Adipocytes are the central cellular component of adipose tissue. However, adipose tissue also has a stromal-vascular fraction containing many immune cells, such as macrophages and T cells. WAT is specialized to store and release lipids, while BAT is mainly involved in thermogenesis. Interestingly, some white adipocytes are able to transform to beige adipocytes displaying a phenotype similar to brown adipocytes. During the development of overweight and obesity, adipocytes first grow in size (hypertrophy) and then in number (hyperplasia) attracting a large number of M1-type macrophages. This induces a change in the adipokine production profile of WAT and leads to chronic inflammation in the tissue (Chap. 7). Genetic variations modulate the individual's susceptibility to obesity in context of the modern obesogenic environment. Studying monogenic forms of obesity provides strong evidence for a central role of appetite regulation in obesity susceptibility with the leptin-melanocortin pathway having an integral role in satiety signaling.

In this chapter, we will define obesity and describe the main functions of adipose tissue and the cells of which the tissue is formed. We will discuss adipogenesis at the example of the conversion of white into beige adipocytes. In this context, we will present the large impact of adipokines during hypertrophy and hyperplasia of WAT and for the communication with the CNS. We will explain how hormonal signals and genetic factors can control food intake and energy expenditure. Variations of the respective genes will be presented as important drivers of monogenetic and common obesity.

Keywords BMI • Visceral fat • Sub-cutaneous fat • White, beige and brown adipocytes • Adipogenesis • Chronic inflammation • M1 and M2 macrophages • Adipokines • Leptin • Hindbrain • Hypothalamus • MC4R • FTO

8.1 Definition of Obesity

No other tissue of the human body can change its dimension as dramatically as the adipose tissue. This is accomplished first by increasing the size of individual cells up to a critical threshold (hypertrophy) and then increasing the number by recruiting new adipocytes from the resident pool of progenitors (hyperplasia). The *WHO* defines overweight and obesity as "abnormal or excessive fat accumulation that may

C. Carlberg et al., *Nutrigenomics*, DOI 10.1007/978-3-319-30415-1_8

impair health". Obesity is the consequence of excess WAT accumulation and develops when energy intake exceeds energy expenditure. The most commonly used measure of obesity is the BMI (Sect. 1.2). A person is defined as normal weight if his/her BMI is 18.5–24.9 kg/m^2, overweight if the BMI is 25–29.9 or obese if the BMI is 30 or above (Fig. 8.1a). Individuals with adult-onset of obesity mostly

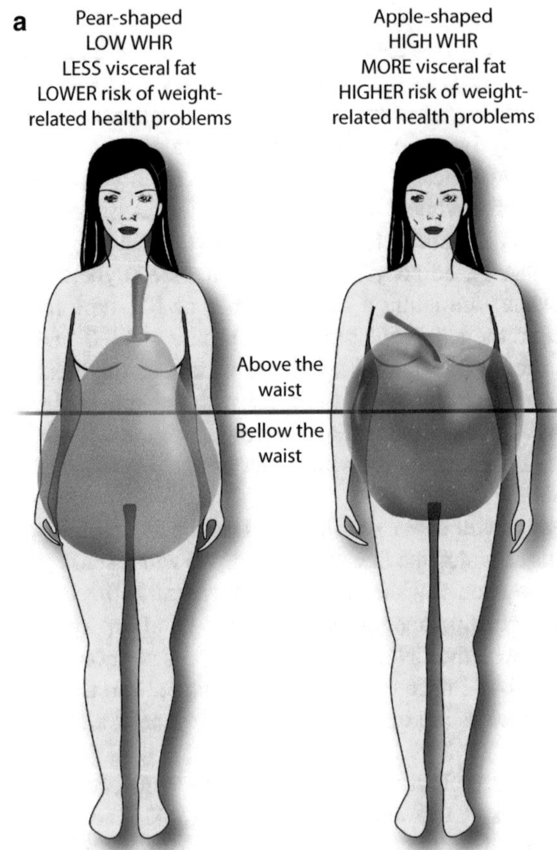

Fig. 8.1 Fat distribution influences obesity associated risks in humans. (**a**). Obesity is defined by a BMI of ≥ 30 and in general is a consequence of fat accumulation. The respective fat distribution in the body can also be measured using anthropometric measures, such as waist circumference or waist-to-hip ratio (WHR). Obese subjects with a low WHR, characterized as "pear-shaped" obesity with predominantly increased sub-cutaneous fat, have a low risk of diabetes and metabolic syndrome. In contrast, obese subjects with a high WHR, characterized as "apple-shaped" obesity with increased visceral fat, have a high risk of for these diseases. (**b**). In humans, WAT is found in all areas of the body. The sub-cutaneous and the intra-abdominal depots are the main fat storage compartments. BAT is abundant at birth and is still present in adults, but to a lesser extent

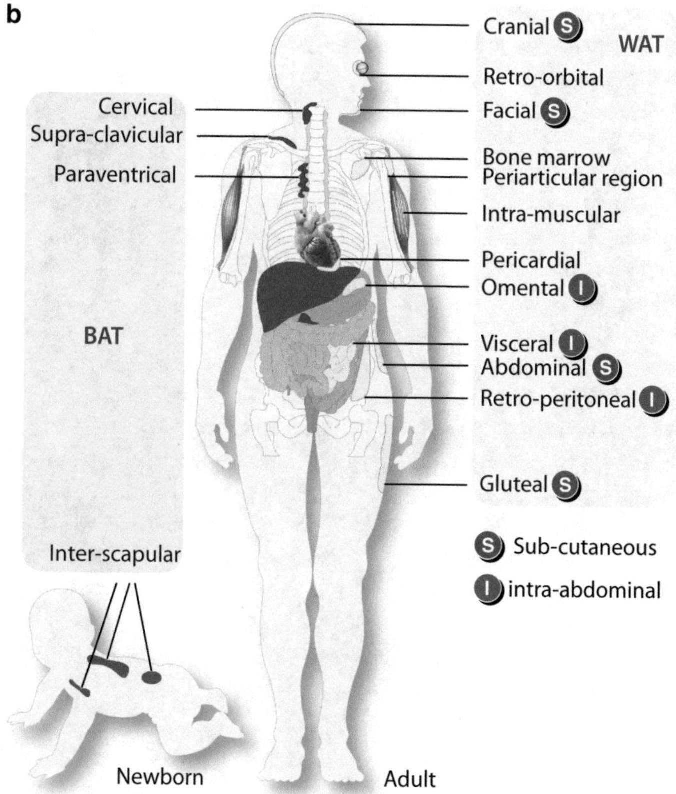

Fig. 8.1 (continued)

exhibit increased adipocyte size, whereas persons with early-onset obesity show both adipocyte hypertrophy and hyperplasia.

The location of WAT in the body also plays an important role for the risk to develop the metabolic syndrome (Chap. 12). High amount of visceral fat (i.e. intra-abdominal fat), referred to as central or "apple-shaped" obesity, increases the risk, while rise in sub-cutaneous fat, referred to as peripheral or "pear-shaped" obesity, exerts far less risk. Obesity is rare to occur in the wild life, but there are examples of animals living in harsh climates, such as polar bears and seals, that are obese. This indicates that a high degree of natural adiposity can even contribute to evolutionary fitness. However, obesity found in modern humans is mostly accompanied by inflammation (Sect. 8.6) and consecutively often by the different features of the metabolic syndrome (Chap. 12). In fact, most of the world's population today lives in countries where individuals are more likely to die from the consequences of being obese than starving (Box 8.1). Of note, human obesity does not always result in disease suggesting that the threshold for tolerable BMI differs among individuals and may be determined by environmental and genetic variables (Sect. 8.7).

Box 8.1 Worldwide Increase of Overweight and Obesity

Between 1980 and 2013, the proportion of overweight or obese adults, i.e. with a BMI of 25 or higher, increased worldwide from 28.8 to 36.9 % in males and from 29.8 to 38.0 % in females. Importantly, in 2013, in developed countries 23.8 % of boys and 22.6 % of girls in the age of 2–19 were overweight or obese, while in developing countries the respective numbers were 12.9 % and 13.4 %. This indicates that within some 30 years the worldwide prevalence of overweight and obesity rose by 27.5 % for adults and even by 47.1 % for children. In absolute numbers this represents an increase from 857 million in 1980 to 2.1 billion in 2013. Since 2006 the dynamics in adult obesity has slowed down in developed countries, while the prevalence of obesity exceeded 50 % in men in Tonga and in women in Kuwait, Kiribati, Micronesia, Libya, Qatar, Tonga and Samoa. However, at all ages, prevalence of overweight and obesity was higher in developed than in developing countries (Fig. 8.2). In developed countries, males older than 10 years showed higher rates of overweight and obesity than females, while in developing countries, women have higher rates than men older than 25 years. However, 62 % of the worldwide 671 million obese individuals (2013) live in developing countries. During the last three decades no single country showed a significant decrease in obesity implying the danger that over time most countries are on a trajectory to reach the same high rates as already observed in Tonga or Kuwait.

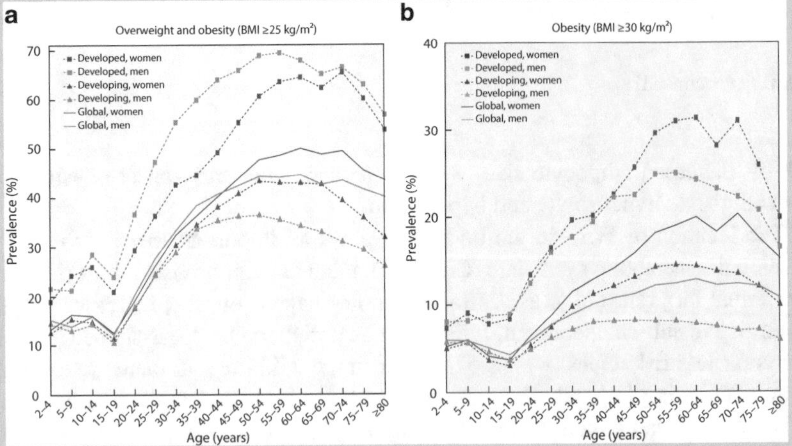

Fig. 8.2 Age pattern of overweight and obesity in 2013. Worldwide prevalence of overweight and obesity (**a**) and obesity alone (**b**) is shown by age and sex in developed and developing countries (data collected from 188 countries by Ng et al. Lancet *384*, 766–781)

8.2 Adipogenesis

Storing excess fat in vesicles is an evolutionary adapting process, which already started in worms, such as *C. elegans*. Most vertebrate species store fat in a tissue of a mesodermal origin, named WAT. In humans, WAT is found throughout the body, such as around the omentum, intestines and perirenal areas, as well as subcutaneously in the buttocks, thighs and abdomen (Fig. 8.1b). Moreover, WAT can be found also on the face and extremities and within the bone marrow. In newborns, BAT is found in the neck, kidneys and adrenal regions, while in adults it locates in the neck as well as in supraclavicular and paravertebral regions.

WAT represents the large majority of human adipose tissue and is the primary site of energy storage, while minor amounts are BAT representing a site of basal and inducible energy expenditure. WAT buffers nutrient availability and demand by storing excess calories and prevents toxic lipid levels in non-adipose tissues. This is an essential function, as it allows intervals of fasting between meals and intervals of prolonged week-long fasting. In contrast, BAT maintains core body temperature in response to cold stress by generating heat, i.e. it is primarily used for non-shivering thermogenesis. Both white adipocytes in WAT and brown adipocytes in BAT store triacylglycerols. However, white adipocytes have one large lipid droplet filling 90 % of the cell, while brown adipocytes carry many single lipid compartments and a far larger number of mitochondria than WAT. These mitochondria are enriched with the protein UCP1 that acts as a long-chain fatty acid/H^+ symporter. The action of this protein causes a proton leak across the inner mitochondrial membrane, thus "uncoupling" fuel oxidation from ATP synthesis. Interestingly, also WAT can recruit $UCP1^+$ cells, which are termed beige adipocytes. There are specific WAT depots that can develop high numbers of beige adipocytes, a process referred to as "browning". In reverse, these cells can again increase lipid storage and then morphologically resemble classic white adipocytes, referred to as "whitening". This tissue conversion is an adaptive process, i.e. it depends on environmental challenges, such as low temperatures for browning or a high-fat diet for whitening.

Although there is a significant functional overlap between beige and brown adipocytes, both cell types differ in their gene expression profile. For example, beige adipocytes have a greater inducibility of *UCP1* gene expression and a respectively increased uncoupled respiration. From the three different "shades" of fat, i.e. white, beige and brown, white adipocytes have the major impact on obesity-related metabolic complications, such as the metabolic syndrome (Chap. 12). In rodents, energy expenditure via the BAT contributes to the body weight. Obese persons and T2D patients have lower amounts of BAT suggesting that also in humans the BAT has a role in metabolic diseases. Since beige adipocytes also contribute to heat production, they may counteract obesity and metabolic disease. A possibility to treat and/or prevent accumulation of WAT in obesity and its metabolic complications is to induce browning of WAT via the induction of adipocyte differentiation into the beige phenotype.

Mesenchymal stem cells are the precursors to fat, bone and muscle cells. Growth factors of the bone morphogenetic protein (BMP) and FGF families are central in the first phase of the development of adipocytes. In this commitment phase, the multi-potent mesenchymal stem cells differentiate to WAT and BAT precursors. Both brown adipocytes and myocytes derive from paraxial mesoderm-derived progenitor cells that express the transcription factors myogenic factor 5 (MYF5) and paired box (PAX) 7 (Fig. 8.3). However, in adults brown adipocytes can also develop from skeletal muscle satellite cells. White adipocytes derive from both MYF5$^-$ and MYF5$^+$ progenitors. Beige adipocytes either differentiate from WAT precursor cells or directly from white adipocytes after exposure to cold. Like in other differentiation processes in the human body, transcription factor networks play the main role in driving adipocyte differentiation towards the brown, beige or white phenotype.

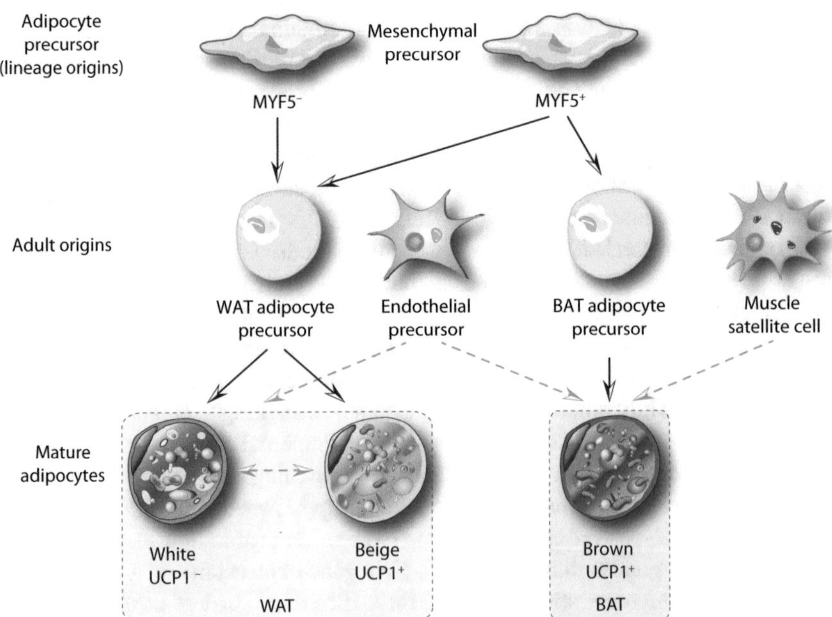

Fig. 8.3 Origins of white, beige and brown adipocytes. BAT contains UCP1 expressing brown adipocytes (UCP1$^+$), whereas WAT is formed by UCP1$^-$ white adipocytes and UCP1$^+$ beige adipocytes. In adults, the expansion of adipose tissue is mainly achieved through the growth and differentiation of pre-adipocytes (i.e. adipocyte precursors). The precursors of WAT and BAT adipocytes derive from mesenchymal cells: for WAT they derive from both MYF5$^+$ and MYF5$^-$ lineages, whereas for BAT they come exclusively from the MYF5$^+$ lineage. Beige adipocytes are obtained from WAT adipocyte precursors or directly from mature white adipocytes. In contrast, brown adipocytes can derive from stem-cell-like skeletal muscle satellite cells. In addition, brown and white adipocytes are generated from endothelial precursors

The most important transcription factors during the differentiation process of white and brown adipocytes are the nuclear receptor PPARγ (Sect. 3.3) and the pioneer factors CEBPA, CEBPB and CEBPD. PPARγ is the "master regulator" of fat cell formation, as it is both necessary and sufficient for adipogenesis. PPARγ can transform non-adipogenic cells, such as fibroblasts and myoblasts, into adipocytes. Interestingly, activation of PPARγ by synthetic ligands potentiates in particular brown adipogenesis and induces a phenotype switch of white into beige adipocytes. Moreover, co-factors of these transcription factors differentially support adipogenesis: nuclear receptor interacting protein 1 (NRIP1) and NCOA2 support the differentiation of white adipocytes, whereas PPARα, FOXC2, PPARGC1A, NCOA1 and PR domain containing 16 (PRDM16) are important for the formation of brown adipocytes. The co-factor PPARGC1A is also involved in the control of mitochondria biogenesis and oxidative metabolism (Sect. 5.2), while PRDM16 is a specific inducer of genes involved in thermogenesis. In addition, like in every differentiation process, epigenetic changes via chromatin modifiers, such as euchromatic histone-lysine N-methyltransferase 1 (EHMT1) or the deacetylase SIRT1 (Sect. 5.4), control also in adipocytes the access of chromatin for transcription factors and their co-factors. Finally, adipogenesis is fine-tuned by the action of miRNAs. During brown adipocyte differentiation, miR-155 forms a feedback loop with CEBPB, while miR-27 acts as a negative regulator of adipogenesis of the brown and beige phenotype and miR-196a expression is increased during the generation of white adipocytes.

White, beige and brown adipocytes can undergo adaptive and dynamic changes in response to starvation or overfeeding as well as in response to cold environment via energy-sensing pathways. One central and illustrative example is the transformation of white adipocytes into beige adipocytes at cold temperature exposure. Some of the signals that regulate this tissue conversion are synthesized locally within the adipose tissue, i.e. they act paracrine. However, other essential factors are of endocrine nature, as they are produced by metabolically highly active organs, such as brain, muscle, heart and liver (Fig. 8.4). In response to exposure to cold temperature, catecholamines are released by the sympathetic nervous system (SNS), but adrenaline can also be secreted by M2-type macrophages in adipose tissue in response to cold stress. This may be counteracted by M1-type macrophages that are present in hypertrophic adipose tissue of obese individuals (Sect. 8.4).

Catecholamine-activated β3-adrenergic receptors, PTGS2-generated prostaglandins and the growth factors BMP4, BMP7 and FGF21 are the key molecules that promote browning of white adipocytes. For example, BMP7 induces the expression of the proteins PRDM16 and PPARGC1A that are central for brown adipocyte differentiation. Moreover, BMPs directly regulate thermogenesis in mature brown adipocytes by increasing their responsiveness to catecholamines and up-regulating intercellular lipase activity via the PKA-MAPK signal transduction pathway. The transformation of white into beige adipocytes is further supported by the growth factors FGF21 and brain-derived neurotrophic factor (BDNF) as well as irisin, which is a peptide hormone produced by skeletal muscle in response to exercise.

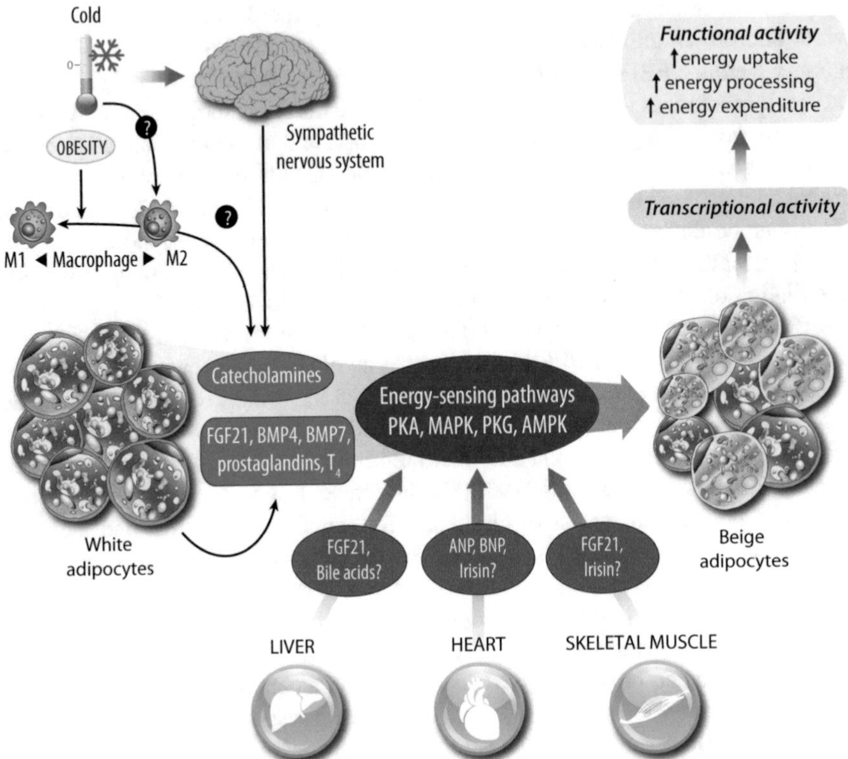

Fig. 8.4 Hormonal control of WAT browning. Metabolic adaptions to environmental factors are regulated by the release of endocrine and paracrine factors from metabolic tissues. In response to (thermal) cold, catecholamines are released by the SNS and from M2-type macrophages in adipose tissue. This activates energy-sensing pathways in white adipocytes and stimulates their transformation to beige adipocytes. This beige phenotype is generated through the actions of transcription factors that induce activities characteristic of their phenotype, such as increase energy uptake, energy processing and energy expenditure

8.3 Energy Homeostasis

Energy homeostasis in adults is achieved by a combination of processes that manage (i) energy intake, (ii) energy storage in form of glycogen in liver, kidney and skeletal muscles and triacylglycerols in WAT and (iii) energy usage, in order to maintain a stable body weight. Therefore, food intake is an integrated response over a prolonged period of time that maintains the levels of energy stored in adipocytes. However, as the result of a daily tiny but cumulative positive energy balance, overweight and obesity can develop in the course of many years. This misbalanced energy homeostasis is multi-factorial and complex (Fig. 8.5). Food consumption has changed radically the last 100–200 years, leading to dramatic changes in our

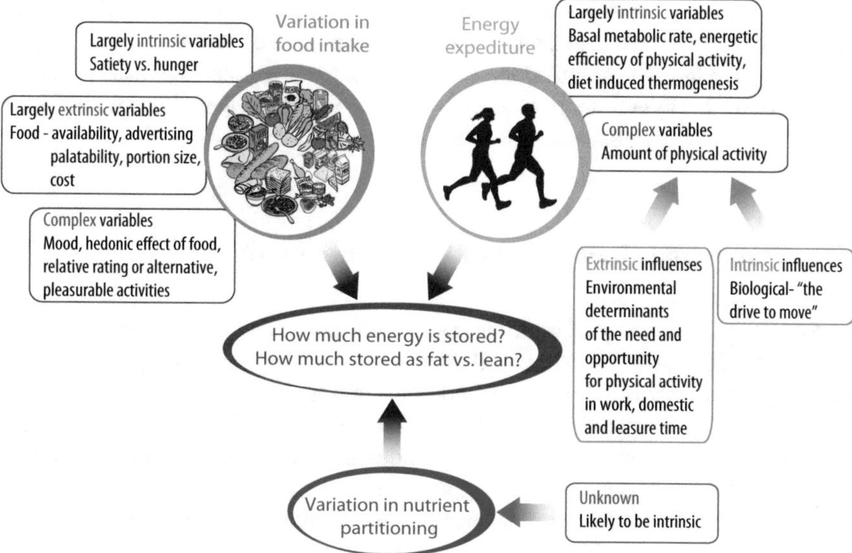

Fig. 8.5 Variables of energy homeostasis. Energy homeostasis is regulated by a complex interaction between the variation in food intake, the tendency to store excess energy as fat or lean mass, referred to as nutrient partitioning, and the variation in energy expenditure. The intrinsic variables that have an impact on energy homeostasis causing obesity are influenced by food intake through effects on satiety and hunger

macronutrient intake. While the contribution of macronutrients in the diet of our ancient ancestors has been estimated as 34 % protein, 45 % carbohydrate and 21 % fat, a typical Western diet today contain 12 % protein, 46 % carbohydrate and 42 % fat. However, this distribution differs between countries. Excess of calorie intake, altered food composition and physical inactivity that are promoted by today's obesogenic environment are the most likely drivers of the obesity epidemic (Sect. 8.1).

The coordinated secretion of numerous hormones from the CNS prepares the digestive system for the anticipated caloric load (Sect. 8.4). Ideally, satiation hormones (being secreted in response to ingested nutrients) control the amount of food intake. In turn, adiposity hormones (indicating the fat content of the body) modify these signals. However, many non-homeostatic factors, such as stress, cultural habits and social influences interact with these hormonal controllers of food intake. In theory, establishing a negative energy balance, i.e. eating less calories than daily consumed by the basal metabolic rate and additional physical activity, would be an easy solution in reducing the problem of overweight and preventing the development of obesity. However, hormonal and neuronal control circuits have been trained by evolutionary adaption to make hunger the first ranking desire of nearly all humans, which strongly counteracts to most attempts losing body weight. In addition, changes in environmental exposures early in life, before, during or after pregnancy, i.e. nutrition-triggered epigenetic programming (Sect. 5.5), cause a sustained long-term effect on the pre-disposition to develop overweight and obesity. Molecular genetic studies have shown that genetic variants causing severe familial obesity largely influence food intake through effects on hunger and satiety (Sect. 8.7). Moreover, the genes that are preferentially influenced by genetic variations associated with adiposity in the general population are pre-dominantly expressed in the brain.

8.4 Hormonal Regulation of Food Anticipation

The coordinated activity of multiple peptide hormones and autonomic circuits are needed for efficient digestion of food. Prior an anticipated meal, the gastrointestinal tract is prepared for the digestion of nutrients and for avoiding extreme metabolic consequences of the pending caloric load. Human individuals who habitually eat at the same time each day begin these CNS-initiated hormone secretions, such as insulin, before food serving. This is important for the efficient disposal of absorbed glucose (Sect. 9.1). Moreover, CNS signals also stimulate ghrelin secretion from the stomach approximately 30 min before the meal, and GLP1 from the intestine rises even 1 h earlier. When food is consumed, numerous hormones and enzymes are secreted, in order to allow nutrient digestion and absorption. Most of these hormones related to digestion, such as cholecystokinin, are satiety signals. Further gastrointestinal peptides that influence the hindbrain to reduce the meal size are GLP1, glucagon, APOA4 and peptide YY. The half-life of these peptides is short and the function of most of them is redundant, i.e. they can compensate each other. Satiation signals converge in the nucleus tractor solitarius (NTS) and the adjacent area postrema of the hindbrain (Fig. 8.6).

Insulin and leptin are obesity signals, i.e. their secretion is proportional to the amount of body fat. Together with ghrelin, they enter the brain through the blood–brain barrier via receptor-mediated active transport and act directly on the arcuate nucleus (ARC) of the hypothalamus (Fig. 8.6). This area of the brain also receives information from the hindbrain on the progress of meals and from limbic centers reflecting non-homeostatic influences. In addition, the peptide hormone amylin that is secreted like insulin from β cells of the pancreas, as well as cytokines derived from adipose tissue, such as IL6 and TNF, all act on the brain to increase energy expenditure. The ARC contains two populations of neurons that either express neuropeptide Y (NPY) and agouti-related peptide (AGRP) or pro-opiomelanocortin (POMC) (Fig. 8.6). POMC is a pro-hormone that is cleaved to produce α-melanocyte-stimulating hormone (α-MSH) being a ligand of the melanocortin 4 receptor (MC4R). In contrast, insulin and leptin activate POMC neurons that release α-MSH counteracting AGRP and stimulating MC4R neurons. This results in reduced food intake and decreased body weight in the long-term inhibition of food intake. Ghrelin stimulates NPY-AGRP neurons that secrete AGRP acting as an antagonist of MC4R on neurons in the paraventricular nuclei (PVN). This leads to increased food intake and induction of weight gain. These areas of the forebrain also receive information

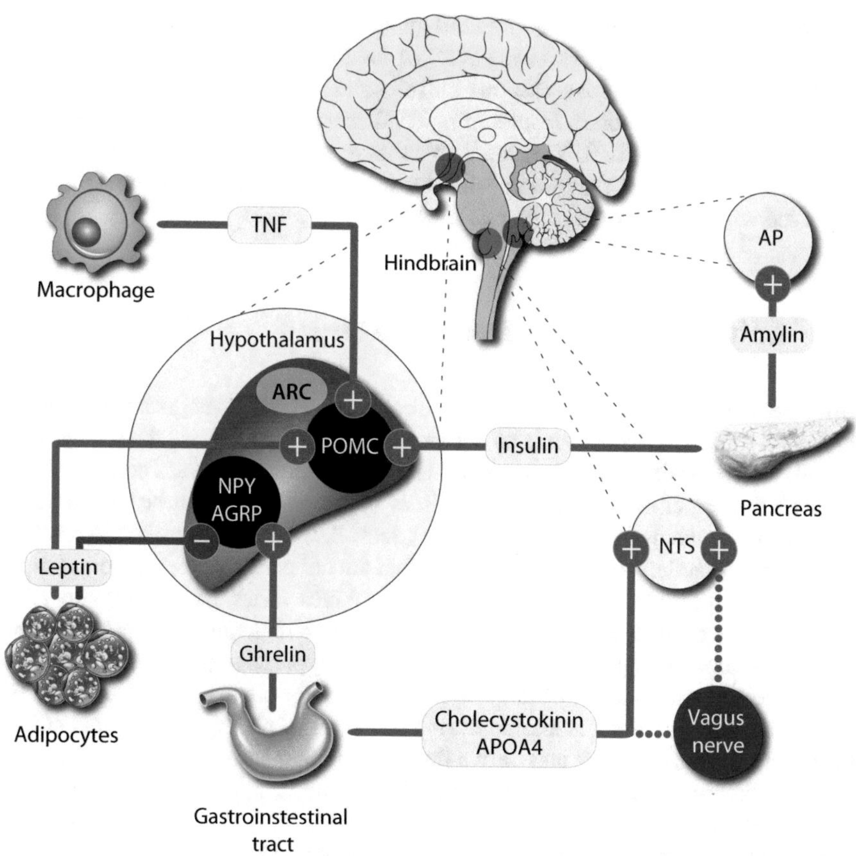

Fig. 8.6 Hormonal signals from the periphery influence multiple brain areas. Insulin and amylin from the pancreas stimulate POMC neurons in the ARC of the hypothalamus and in the AP of the hindbrain, respectively. Ghrelin stimulates NPY-AGRP neurons in the ARC, while leptin from adipocytes inhibits these cells. However, leptin and TNF stimulate POMC cells. Gastrointestinal peptides, such as cholecystokinin and APOA4, either stimulate the NTS directly or stimulate vagal afferent nerves whose axons end in the NTS. Most of these peptide hormones enter the brain via active transport

from the hindbrain concerning the progress of the meal and information from non-homeostatic factors.

Reduced leptin and insulin signaling in the ARC results in increased food intake and eventually weight gain. Also the cytokines IL6 and TNF, derived from adipose tissue, reduce food intake. In particular TNF is associated with the cachexia seen in cancer patients. Adiponectin is secreted from adipose tissue in inverse proportion to body fat and acts on the brain to increase energy expenditure. Some amino acids, such as leucine, glucose and fatty acids, can also regulate the activity of ARC neurons making them important for normal regulation of body weight/fat mass. Insulin and leptin can also regulate ongoing food intake by directly adjusting the sensitivity to incoming satiation signals in the NTS of the hindbrain. For example, when an individual loses weight, the secretion of insulin and leptin decreases and the reduced adiposity signal in the brain results in reduced sensitivity to the satiating action of cholecystokinin and GLP1. This leads to an increase in meal size until lost weight is regained and the levels of adiposity signals are restored. The opposite can be observed when people overeat and gain weight.

8.5 Adipose Tissue is an Endocrine Organ

Adipose tissue is not only a passive storage container for nutrients, as it was believed to be some decades ago, but it is an active endocrine organ that via the reception of signals as well as the secretion of paracrine and endocrine molecules communicates with all important organs of the body. This communication can be mediated by nutritional mechanisms, neural pathways (Sect. 8.4) and via autocrine, paracrine and endocrine actions of secreted proteins that are collectively referred to as adipokines (Table 8.1). The secretion pattern of adipokines varies, based on the location of the adipose tissue within the body, but the two most abundant fat depots, visceral and sub-cutaneous adipose tissue, display a unique expression profile. However, the adipokine secretion profile of adipocytes significantly changes when the cellular composition of the tissue is altered during the onset of obesity. Since most adipokines act pro-inflammatory and only a few anti-inflammatory, their overall expression is increased in the obese state compared to the lean state. Pro-inflammatory adipokines are: the peptide hormones leptin and resistin, the transport proteins retinol binding protein 4 (RBP4) and lipocalin 2, the growth factors angiopoietin-like protein 2 (ANGPTL2), the enzyme NAMPT, the cytokines TNF, IL6 and IL18 and the chemokine (C-X-C motif) ligand 5 (CXCL5). In contrast, only the adipokines adiponectin and secreted frizzled-related protein 5 (SFRP5) act anti-inflammatory. The balance between pro-inflammatory and anti-inflammatory adipokines is crucial for determining homeostasis throughout the body based on the nutritional status.

Leptin is the most important hormone produced by adipose tissue, since it regulates feeding behavior through the CNS (Sect. 8.4). The hormone improves metabolic dysfunction in patients with lipodystrophy or inherited leptin deficiency. However, although obese individuals have high leptin concentrations, they do not show any anorexic response. In addition to its CNS functions, leptin has effects on cells of the immune system. It stimulates the production of pro-inflammatory cytokines and chemokines in monocytes and macrophages, such as TNF, IL6 and CXCL5. Furthermore, leptin polarizes T cells towards a T_H1 phenotype by inducing the production of the T_H1-type cytokines IL2 and IFNγ and the suppression of the T_H2-type cytokine IL4. All these effects confirm that leptin is a pro-inflammatory adipokine. The peptide hormone resistin is also associated with inflammation, since it promotes the expression of the pro-inflammatory cytokines TNF and IL6. At least in mice, resistin induces insulin resistance via the activation of an inhibitor of insulin

Table 8.1 Sources and functions of key adipokines

Adipokine	Primary source(s)	Binding partner or receptor	Function
Adiponectin	Adipocytes	Adiponectin receptors 1 and 2, T-cadherin, calreticulin-CD91	Insulin sensitizer, anti-inflammatory
SFRP5	Adipocytes	WNT5a	Suppression of pro-inflammatory, WNT signaling
Leptin	Adipocytes	Leptin receptor	Appetite control through the central nervous system
Resistin	Peripheral blood mononuclear cells (PBMCs)	Unknown	Promotes insulin resistance and inflammation through IL-6 and TNF secretion from macrophages
RBP4	Liver, adipocytes, macrophages	Retinol (vitamin A), transthyretin	Implicated in systemic insulin resistance
Lipocalin 2	Adipocytes, macrophages	Unknown	Promotes insulin resistance and inflammation through TNF secretion from adipocytes
ANGPTL2	Adipocytes, other cells	Unknown	Local and vascular inflammation
TNF	Stromal vascular fraction cells, adipocytes	TNF receptor	Inflammation, antagonism of insulin signaling
IL6	Adipocytes, stromal vascular fraction cells, liver, muscle	IL6 receptor	Changes with source and target tissue
IL18	Stromal vascular fraction cells	IL18 receptor	IL18 binding protein, broad-spectrum inflammation
CCL2	Adipocytes, stromal vascular fraction cells	CCR2	Monocyte recruitment
CXCL5	Stromal vascular fraction cells (macrophages)	CXCR2	Antagonism of insulin signaling through the JAK-STAT pathway
NAMPT	Adipocytes, macrophages, other cells	Unknown	Monocyte chemotactic activity

signaling, the protein suppressor of cytokine signaling 3 (SOCS3). Moreover, resistin directly counteracts the anti-inflammatory effects of adiponectin on vascular endothelial cells.

The main source of RBP4 is the liver, but also adipocytes and macrophages can produce the transporter of vitamin A (retinol). In an auto- or paracrine manner RBP4 inhibits insulin-induced phosphorylation of IRS1, i.e. the adipokine is involved in the regulation of glucose homeostasis in adipocytes. Increased serum RBP4 levels are associated with high blood pressure, low HDL-cholesterol levels, high levels of LDL-cholesterol and triacylglycerols and increased BMI. Lipocalin 2 belongs to the same protein superfamily as RBP4 and transports various small lipophilic substances, such as retinoids, arachidonic acid and steroids. The protein is induced by inflammatory stimuli through the activation of NF-κB and high lipocalin 2 concentrations are found in obese individuals. Serum levels of lipocalin 2 are associated with adiposity, hyperglycemia, insulin resistance and CRP concentrations.

ANGPTL2 is a growth factor that induces inflammatory responses and activates integrin signaling in endothelial cells, monocytes and macrophages. This protein can induce insulin resistance and its serum levels are associated with obesity, insulin resistance and CRP concentrations. The enzyme NAMPT, also called visfatin, is mainly expressed and secreted by adipose tissues. High circulating levels of NAMPT are found in obesity and in T2D patients. NAMPT expression correlates with the serum levels of IL6 and CRP. NAMPT is essential for the biosynthesis of NAD and has an important role in controlling the insulin secretion of β cells (Sect. 3.6).

TNF is a pro-inflammatory cytokine with a prominent role in basically all inflammatory and autoimmune diseases. The cytokine is mainly produced by monocytes and macrophages, but it can also be secreted by activated adipocytes. TNF promotes insulin resistance in skeletal muscle and adipose tissues by reducing the phosphorylation of insulin receptor (IR) and IRS1 (Sect. 9.3). TNF concentrations are increased in adipose tissue and plasma of obese individuals. Like TNF, also IL6 is a pro-inflammatory cytokine that is involved in obesity-related insulin resistance. Serum IL6 levels correlate with adiposity and are increased in T2D patients. Adipose tissue is a major source of the cytokine, since more than 30 % of all circulating IL6 is produced there. IL6 shows disparate actions on insulin signaling in different organs, such as liver versus muscle. This is due to the different sources of IL6, such as muscle versus adipose tissue. Another pro-inflammatory cytokine produced by adipose tissues is IL18. In obese individuals IL18 serum levels are increased, but they decline following weight loss. Atherosclerotic lesions show high IL18 levels and indicate plaque instability. The chemokine CXCL5 is secreted by macrophages within the stromal vascular fraction of adipose tissue and is associated with inflammation and insulin resistance. In obese insulin-resistant individuals, circulating levels of CXCL5 are higher than in obese insulin-sensitive individuals. CXCL5 interferes with insulin signaling in muscles by activating the Janus kinase (JAK)-signal transducer and activator of transcription (STAT) pathway through its receptor CXC-chemokine receptor 2 (CXCR2). TNF controls *CXCL5* gene expression via the activation of NF-κB.

The peptide hormone adiponectin is exclusively synthesized by adipocytes and is found at high levels in serum. Compared with lean individuals, obese individuals have decreased adiponectin levels. Adiponectin expression in adipocytes is inhibited by pro-inflammatory cytokines, such as TNF and IL6, as well as by hypoxia and oxidative stress, respectively. In contrast, adiponectin secretion is stimulated by PPARγ during adipocyte differentiation. Moreover, plasma adiponectin levels inversely correlate with visceral fat accumulation and are decreased in T2D patients, while high adiponectin levels are associated with a lower risk for developing T2D. Adiponectin is the adipokine that is expressed at the highest levels in functional adipocytes of lean persons, while its expression is down-regulated in dysfunctional adipocytes of obese individuals. The beneficial effects of adiponectin on insulin sensitivity are mediated via increased Ca^{2+} levels in skeletal muscle that activate CAMKK2, AMPK and SIRT1 and result in the up-regulation of *PPARGC1A* expression. Adiponectin also modulates the function and phenotype of macrophages. For example, it inhibits the transformation of macrophages into foam cells (Chap. 11.2) by reducing the intracellular cholesteryl ester content and suppressing the expression of macrophage scavenger receptor 1 (MSR1). Furthermore, in macrophages adiponectin up-regulates the secretion of the anti-inflammatory cytokine IL10. Taken together, adiponectin induces a shift from M1-type to M2-type macrophages.

The main function of the other anti-inflammatory adipokine, SFRP5, is to prevent the binding of wingless-type MMTV integration site family member (WNT) proteins to their respective receptors. The WNT signaling pathway has a number of important downstream targets, such as MAPK8, leading to pro-inflammatory cytokine production in macrophages. WNT5A antagonizes SFRP5 and the balance between SFRP5 and WNT5A in adipocytes and adipose tissue macrophages, modulates inflammation and metabolic function.

Taken together, adipose tissue influences and communicates via adipokines with many other organs, such as the brain, heart, liver and skeletal muscle, and vascularization, respectively. During adipose tissue expansion adipocyte dysfunction often occurs, such as a dys-regulation of adipokine production, which has both local and systemic effects on inflammatory responses. This changed adipokine profile significantly contributes to the initiation and progression of obesity-induced cardiovascular and metabolic diseases (Chaps. 11 and 12).

8.6 Inflammation in Adipose Tissue

In WAT, lipid-loaded adipocytes represent only 20–40 % of the cell number of a fat pad but more than 90 % of its volume, i.e. every gram of adipose tissue contains 1–2 million adipocytes but 4–6 million stromal-vascular cells, of which more than half are immune cells. As already discussed in Sect. 7.4, obesity causes the recruitment and infiltration of M1-type macrophages to WAT, which is the initial event in obesity-induced inflammation. These macrophages secrete a variety of pro-inflammatory cytokines, which act locally in a paracrine manner or they leak out of the adipose tissue and cause systemic effects, such as insulin resistance in metabolic organs (Sect. 9.4). The inflammatory response in obesity is chronic, i.e. it persists far longer than acute inflammation caused by infections or tissue injury.

Adipose tissue can be classified into at least three structural and functional groups. Lean individuals with normal metabolic function store excess nutrients as triacylglycerols in WAT. The WAT in these lean subjects contains M2-type macrophages and T_H2 cells that respond to nutrient-derived signals by promoting lipid storage and suppressing lipolysis (Fig. 8.7, stage 1). As obesity develops with chronic overnutrition, the storage capacity is exceeded causing cellular dysfunction, such as lipid dys-regulation, mitochondrial dysfunction, oxidative stress and ER stress (Sect. 7.5), leading to reduced metabolic control. This causes adipocytes to secrete chemokines, such as CCL2, that attract monocytes into the adipose tissue that become M1-type macrophages (Fig. 8.7, stage 2). When these alterations escalate, they lead to adipocyte death. In order to remove remnants of dead adipocytes, additional macrophages infiltrate the WAT. They surround the dead cells and create crown-like structures that are associated with increased inflammation (Fig. 8.7, stage 3). T cells in adipose tissue also play a role in obesity-induced inflammation. T_H1 cells produce pro-inflammatory cytokines, such as IFNγ, while T_H2 cells and T_{REG} cells secrete anti-inflammatory cytokines, such as IL10, inducing differentiation of macrophages into M2 type (Sect. 7.2). In lean individuals, however, T_H2 and T_{REG} cells dominate in WAT, while in obese persons there are far more T_H1 cells. Compared to sub-cutaneous fat, visceral fat accumulates a larger number of macrophages and secretes greater amounts of pro-inflammatory cytokines. In addition, adipocytes in visceral fat are more fragile and reach earlier a critical size triggering cell death than sub-cutaneous adipocytes. This may explain the different health risk between the two "apple" and "pear" shapes of fat depots (Sect. 8.1).

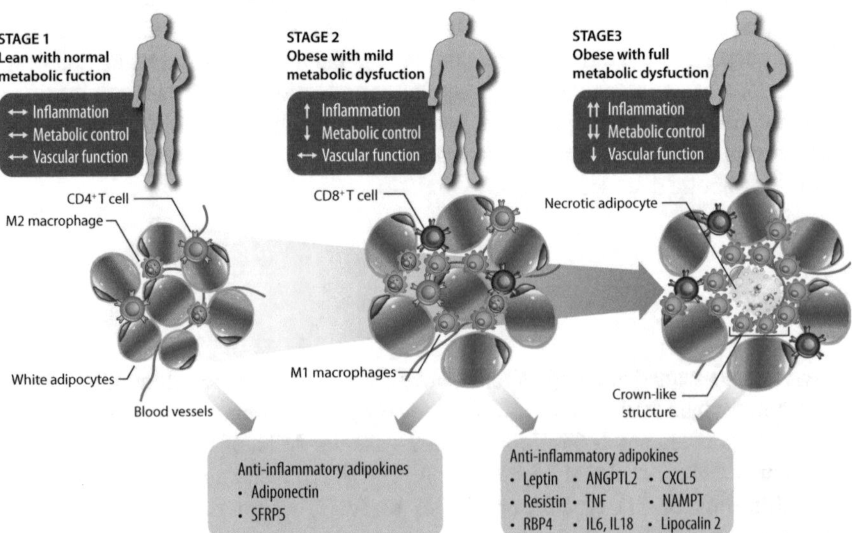

Fig. 8.7 Functional classification of adipose tissue. Adipose tissue can be distinguished into at least three stages. In normal-weight tissue with normal metabolic function (stage 1) adipocytes are associated with a rather low number of M2-type macrophages. This tissue produces preferentially anti-inflammatory cytokines, such as adiponectin and SFRP5. During onset of obesity, adipocytes increase their triglyceride storage, i.e. they become hypertrophic. At limited obesity (stage 2) adipocytes still retain relatively normal metabolic function and display low levels of immune cell activation and sufficient vascular function. However, in obesity with full metabolic dysfunction (stage 3) the tissue has recruited a large number of M1-type macrophages and produces preferentially pro-inflammatory adipokines, such as leptin, resistin, RBP4, lipocalin, ANGPTL2, NAMPT, TNF, IL6, IL18 and CXCL5

8.7 Genetics of Obesity

Some rare forms of severe obesity result from mutations in an individual gene or chromosomal region, i.e. they represent monogenic obesity (Table 8.2). The importance of the leptin-melanocortin pathway in hyperphagia (increased appetite) and obesity susceptibility is indicated by the fact that so far primarily mutations in the genes for leptin (*LEP*), *LEPR*, *POMC*, proprotein convertase subtilisin/kexin type 1 (*PCSK1*), *MC4R* and single-minded family bHLH transcription factor 1 (*SIM1*) were found as causes of monogenetic obesity. *PCSK1* encodes for an enzyme responsible for post-translational processing of POMC, while *SIM1* encodes for a transcription factor that is both an upstream and downstream target of MC4R. *MC4R* mutations may be responsible for up to 6 % of childhood obesity and 2 % of adult obesity cases. Importantly, the very rare obesity phenotype of patients with homozygous *LEP* mutations can be reversed by administration of leptin.

The rapid increase in the number of obese people pointed out above (Fig. 8.2) can be explained by radical changes in lifestyle, such as high intake of energy-dense food and physical inactivity. However, some subjects seem to be more susceptible to these lifestyle changes than others, suggesting a relevant genetic component. Polygenic "common" obesity results from the combined effect of multiple genetic variants in concert with environmental risk factors. Linkage analysis, candidate gene approaches and in particular GWAS (Sect. 2.5) in various populations have indicated dozens of genes to be associated with the traits BMI and obesity (Fig. 8.8). Widely replicated candidate genes are *MC4R*, *BDNF*, *PCSK1*, adrenoceptor beta 3 (*ADRB3*) and *PPARG*. The most prominent result from GWAS analysis was the identification of a strong association of the chromosomal region of the *FTO* gene with BMI and obesity. Although the effect size of the genetic variations at the *FTO* locus is not comparable to that of monogenic forms, it represents the most established association with common obesity, primarily due to its high frequency (47 %) in the European population. The FTO protein shares motifs with iron- and α-ketoglutarate-dependent oxygenases that are involved in fatty acid metabolism and DNA repair, but its exact function is still unknown. FTO is located in the nucleus, may be involved in DNA demethylation and is expressed in multiple tissues, including the hypothalamus. Carriers of *FTO* risk SNPs show increased appetite and measured food intake suggesting that the mechanism of this common genetic variant is, like for rare variants, through increased energy intake.

Further prominent genes identified by GWAS are the transmembrane protein 18 (*TMEM18*) and the ER transporter SEC16 homolog B (*SEC16B*). In total, a recent GWAS meta-analysis of nearly 340,000 individuals identified 97 genomic loci associated with BMI. However, these loci in total account only for 2.7 % of the variation in BMI. The same study suggests that maximal 21 % of the BMI variation may be explained by common genetic variations. Also, in view of this very large study, the role of the CNS is confirmed, in particular of genes expressed in the hypothalamus, in the regulation of body mass. This suggests that obesity may be considered as a "neurobehavioral" disorder with high susceptibility to the obesogenic environment. Taken together, some 100 genes are associated with obesity and fat distribution, but

Table 8.2 Monogenic cases of obesity

Gene	Genomic position	Mode of inheritance	Associated phenotype
Proopiomelanocortin (*POMC*)	2p23.3	Autosomal recessive	Severe pediatric-onset obesity
			Hyperphagia
			Red hair pigmentation
			Pale skin
Leptin (*LEP*)	7q32.1	Autosomal recessive	Severe early-onset obesity
			Extreme hyperphagia
			Hyperinsulinemia
			Hypothalamic hypothyroidism
			Hypogonadotropic hypogonadism
Leptin receptor (*LEPR*)	1p31.3	Autosomal recessive	Severe obesity with hyperphagia
			Delayed or absent puberty
			Reduced IGF1 levels
			Growth abnormalities
Proprotein convertase subtilisin/kexin type 1 (*PCSK1*)	5q15	Autosomal recessive	Severe childhood obesity
			Abnormal glucose homeostasis
			Reduced plasma insulin with elevated proinsulin levels
			Hypogonadotropic hypogonadism
			Hypocortisolemia
Melanocortin 4 receptor (*MC4R*)	18q21.32	Autosomal dominant/ recessive	Severe early-onset obesity
			Hyperphagia
			Highly elevated plasma insulin concentrations
			Increased BMD
Single-minded family BHLH transcription factor 1 (*SIM1*)	6q16.3	Autosomal dominant	Early-onset obesity
			Hypotonia
			Developmental delay
			Short extremities

most of them are not yet mechanistically understood. The small effect size of these common variants implies that they do not have much predictive value. In addition, the genetics of common obesity has a very heterogeneous etiology, but it links with several related metabolic diseases and processes, such as cardiometabolic traits.

Future View

In future, the development and pathogenesis of obesity might be seen as a response to metabolic toxicity. The metabolic syndrome (Chap. 12) may be improved when the nutrient storage capacity of WAT would be improved and the energy-dissipating potential of thermogenic brown and beige adipocytes would be maximized. This

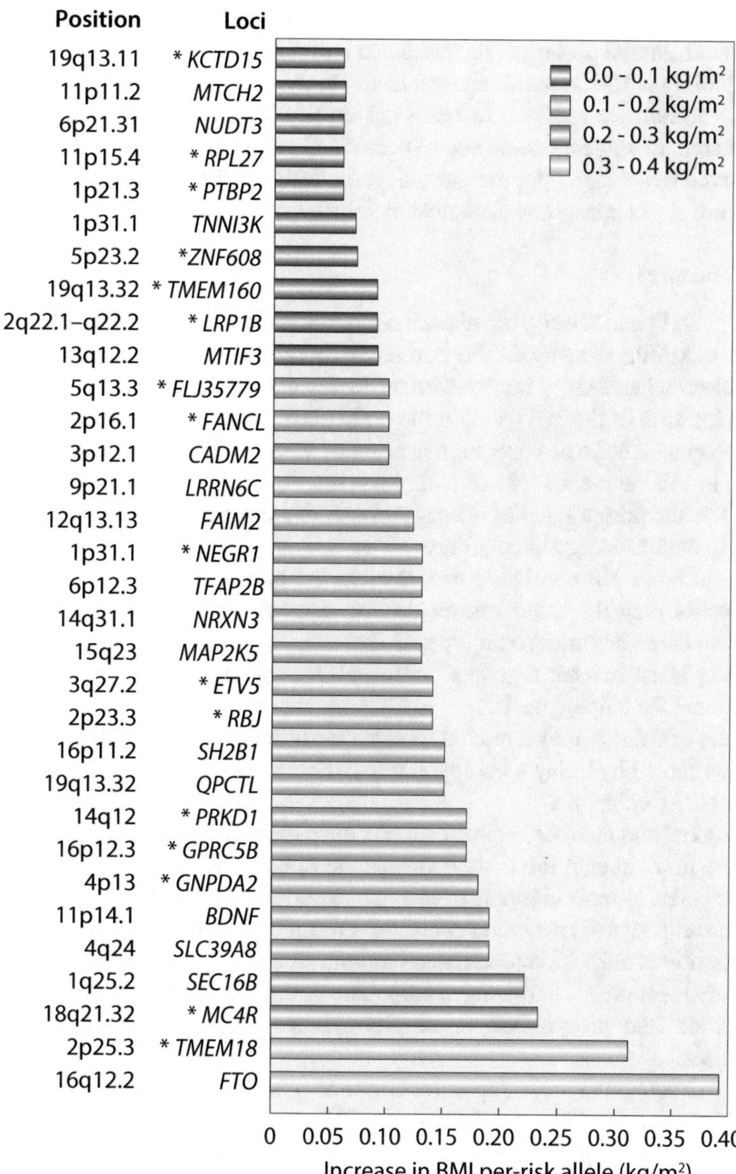

Fig. 8.8 The genetics of BMI. The chromosomal position of genes is listed that are located at or close to (*) SNPs associating with BMI

will include the need of a better understanding of the development and activation of white, beige and brown adipocytes and the role of nutrients on the differentiation of M1- and M2-type macrophages. This may be achieved by insight into the functions and mechanisms of key adipokines and the development of therapeutic strategies counteracting the imbalance of pro-inflammatory and anti-inflammatory adipokines. A substantial portion of the predicted heritability of obesity and inter-individual variability in BMI remains unexplained. Therefore, it can be questioned, whether genetic variations are still the key cause of obesity. Most likely the concept needs to be extended by epigenetic and social-behavioral components. This will include the use of even larger study populations, cleverly designed intervention studies and the development of appropriate statistical methods.

Key Concepts

- Overweight and obesity are abnormal and excessive fat accumulations that may impair health. Obesity is the consequence of excess WAT accumulation and develops when energy intake exceeds energy expenditure.
- The location of the WAT within the body plays an important role for the risk to develop metabolic disease: high amount of visceral fat increases the risk, while gain in sub-cutaneous fat exerts far less risk.
- WAT is the primary site of energy storage, while BAT represents a site of basal and inducible energy expenditure.
- Like in other differentiation processes in the human body, transcription factor networks play the main role in driving adipocyte differentiation towards the brown, beige or white phenotype.
- PPARγ is the "master regulator" of fat cell formation, as it is both necessary and sufficient for adipogenesis.
- Excess of calorie intake, altered food composition and physical inactivity, which are promoted by today's obesogenic environment, are the most likely drivers of the obesity epidemic.
- Hormonal and neuronal control circuits have been trained by evolutionary adaption to make hunger the first ranking desire of nearly all humans, which strongly counteracts to most attempts of losing excessive body weight.
- When adipocyte dysfunction occurs as a result of adipose tissue expansion, dysregulation of adipokine production has both local and systemic effects on inflammatory responses. This changed adipokine profile significantly contributes to the initiation and progression of obesity-induced metabolic and cardiovascular diseases.
- Leptin may be the most important hormone produced by adipose tissue, since it regulates feeding behavior through the CNS.
- To maintain energy homeostasis, the CNS receives hormonal signals from adipose tissue (leptin and other adipokines), from the pancreas and from other parts of the gastrointestinal tract.
- Obesity causes a recruitment and infiltration of M1-type macrophages, which comprise up to 40 % of the cells in the obese adipose tissue.
- The balance between pro-inflammatory and anti-inflammatory adipokines is crucial for determining homeostasis throughout the body based on the nutritional status.
- The importance of the leptin-melanocortin pathway in hyperphagia and obesity susceptibility is indicated by the fact that so far primarily mutations in the genes

LEP, *LEPR*, *POMC*, *PCSK1*, *MC4R* and *SIM1* were found as causes of monogenetic obesity.
- The most prominent result from GWAS analysis was the identification of a strong association of the chromosomal region of the *FTO* gene with BMI and obesity.
- Some 100 genes are associated with obesity and fat distribution, but most of them are not mechanistically understood. The small effect size of these common variants implies that they do not have much predictive value.
- Obesity may be considered as a neurobehavioral disorder with high susceptibility to the obesogenic environment.

Additional Reading

Bartelt A, Heeren J (2014) Adipose tissue browning and metabolic health. Nat Rev Endocrinol 10:24–36

Begg DP, Woods SC (2013) The endocrinology of food intake. Nat Rev Endocrinol 9:584–597

El-Sayed Moustafa JS, Froguel P (2013) From obesity genetics to the future of personalized obesity therapy. Nat Rev Endocrinol 9:402–413

Ng M, Fleming T, Robinson M, Thomson B, Graetz N, Margono C, Mullany EC, Biryukov S, Abbafati C, Abera SF et al (2014) Global, regional, and national prevalence of overweight and obesity in children and adults during 1980–2013: a systematic analysis for the Global Burden of Disease Study 2013. Lancet 384:766–781

Ouchi N, Parker JL, Lugus JJ, Walsh K (2011) Adipokines in inflammation and metabolic disease. Nat Rev Immunol 11:85–97

Peirce V, Carobbio S, Vidal-Puig A (2014) The different shades of fat. Nature 510:76–83

Rosen ED, Spiegelman BM (2014) What we talk about when we talk about fat. Cell 156:20–44

Chapter 9
Glucose Homeostasis, Insulin Resistance and β Cell Failure

Abstract Humans rely on that their blood glucose levels maintain within a physiological range of 4–6 mM during the day with periods of fasting and refeeding. Therefore, glucose intake, storage, mobilization and breakdown are tightly regulated, in particular by the peptide hormones insulin and glucagon. The insulin signaling axis is composed of a number of critical nodes including the insulin receptor (IR), the adaptor protein IRS, the kinases PI3K and AKT as well as the transcription factor forkhead box O (FOXO). Each of these nodes is represented by several protein isoforms and is interconnected with a number of other signal transduction cascades. For example, multiple upstream pathways regulate FOXO activity through post-translational modifications and nuclear-cytoplasmic shuttling of both the transcription factor and its co-regulators. In this way, insulin is controlling a number of physiological functions, but is also sensible to several cellular processes that, when misbalanced, affect insulin signaling, what might lead to insulin resistance. In insulin resistance, normal concentrations of insulin cause an insufficient response of the major insulin target tissues, such as skeletal muscle, liver and adipose tissue. Three main processes can lead to insulin resistance in skeletal muscle and liver: an ectopic overload of lipids, a chronic inflammatory response and ER stress. Similarly, glucotoxic and lipotoxic stress to β cells are mediated via inflammatory response, oxidative stress and ER stress eventually resulting in the failure of the cells.

In this chapter, we will describe the molecular principles of glucose homeostasis and insulin signaling. Using the example of FOXO transcription factors we will analyze the mechanisms how a central signal transduction cascade interacts with environmental challenges mediated via multiple other pathways, in order to keep cells and tissues in homeostasis. This mechanistic understanding will help to integrate the complexity of insulin resistance. We will summarize our insight on chronic inflammation (Sect. 7.2) and the unfolded protein response (Sect. 7.5), in order to interpret the response of muscle and liver cells while developing insulin resistance and that of β cells in the process towards their failure.

Keywords Glucose homeostasis • IR • IRS • PI3K • AKT • FOXO1 • Insulin resistance • Chronic inflammation • Ectopic lipid deposition • MAPK8 • Glucotoxicity • Lipotoxicity • ER stress • Oxidative stress • Unfolded protein response • β cell failure • apoptosis

C. Carlberg et al., *Nutrigenomics*, DOI 10.1007/978-3-319-30415-1_9

9.1 Glucose Homeostasis in Health

Glucose homeostasis results both from the hormonal and neural control of glucose production and use, which even at physiological challenges, such as food ingestion, fasting and intense physical exercise, maintains the blood glucose level within a range of 4–6 mM. This constant level is essential for providing energy to tissues, most importantly for an uninterrupted glucose supply to the brain, which almost exclusively uses glucose as an energy source. Hypoglycemia, i.e. a blood glucose level below 4 mM, can lead in the brain to a number of neuroglycopenic effects, while chronic hyperglycemia, a constant concentration above 10 mM, is toxic to blood vessels (Box 10.1). The principal regulators of glucose homeostasis are the peptide hormones glucagon and insulin that are secreted by α and β cells, respectively, of the endocrine pancreas (forming Langerhans islets). Glucagon is secreted when blood glucose concentrations are low, such as between meals and during exercise. Glucagon has the greatest effect on the liver, where it stimulates (i) the release of glucose that was stored in form of glycogen into the blood and (ii) the production of glucose via the gluconeogenesis pathway.

In contrast, rising blood glucose levels directly after food ingestion stimulates insulin secretion. The glucose transporter GLUT2 in the plasma membrane of β cells and the hexokinase GCK in the cytoplasm both sense glucose (Sect. 3.1) and initiate glucose import and its metabolism via glycolysis (Fig. 9.1). The increasing ATP-ADP ratio stimulates the closing of ATP-sensitive K^+ (K^{ATP}) channels, plasma membrane depolarization, activation of voltage-gated Ca^{2+} channels and Ca^{2+}-mediated stimulation of exocytosis of insulin granules. This K^{ATP} channel-dependent mechanism is a triggering signal that is particularly important for the acute phase of insulin release, i.e. during the first 10 min after glucose rise. In the second phase of insulin secretion, the mitochondrial glucose metabolism generates signals additional to the ATP-ADP ratio, which are important for gaining insight into the functional failure of β cells during T2D. Pyruvate, the final product of glycolysis, flows into mitochondria through an anaplerotic (meaning "refilling") process via pyruvate carboxylase (PC) and an oxidative pathway via the pyruvate dehydrogenase (PDH) complex (Fig. 9.1). The conversion of pyruvate to oxaloacetate via the action of PC and the following metabolism of oxaloacetate to malate, citrate or isocitrate in the TCA cycle provides several possibilities for the reconversion of these metabolites into pyruvate via cytosolic and mitochondrial pathways. These pathways are important for the regulation of glucose-stimulated insulin secretion. One of these pathways is the export of citrate from the mitochondria through the citrate-isocitrate carrier (SLC25A1) and the subsequent conversion of isocitrate to α-ketoglutarate by the cytosolic NADP-dependent IDH complex. The metabolic byproducts of this pyruvate-isocitrate cycling may also act as amplifying signals for the control of glucose-stimulated insulin secretion.

After food ingestion, carbohydrates are digested in the gastrointestinal tract and glucose is absorbed into the circulation primarily via the hepatic portal vein. Thus,

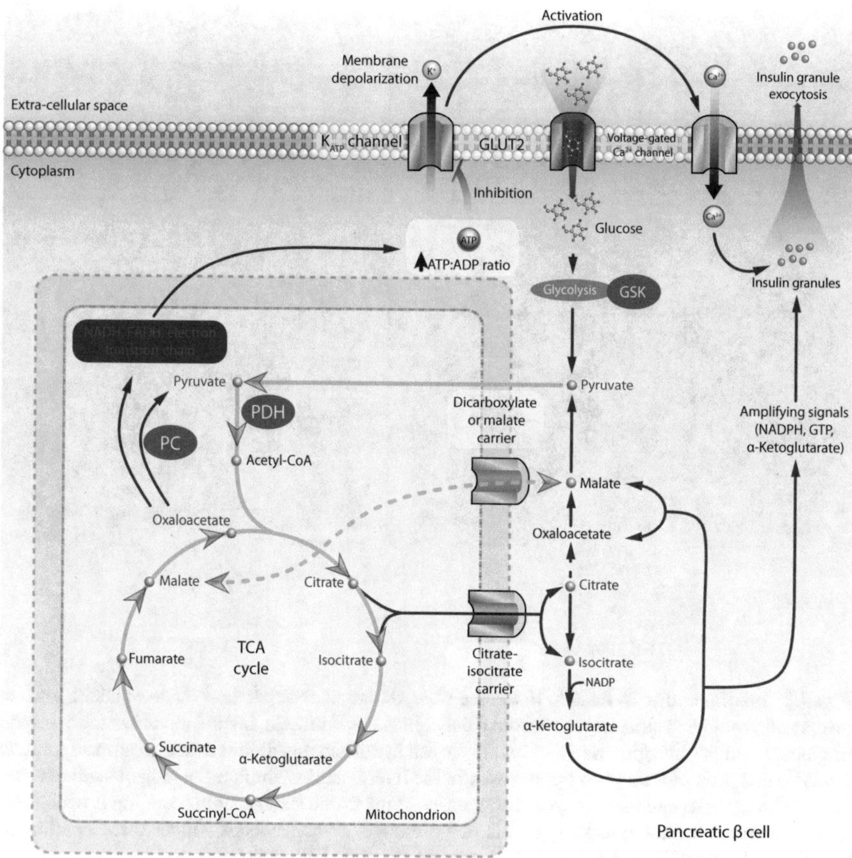

Fig. 9.1 Glucose-stimulated insulin secretion in β cells. Rising blood glucose levels stimulate GLUT2 in the membrane of β cells to import glucose and GCK to start glucose breakdown via glycolysis generating ATP. The increased ATP-ADP ratio inhibits K^{ATP} channels resulting in membrane depolarization, activation of voltage-gated Ca^{2+} channels, influx of Ca^{2+} and stimulation of insulin granule exocytosis. Moreover, pyruvate is the end product of glycolysis and enters mitochondrial metabolism via PDH or PC. The β cells also exhibit active "pyruvate cycling" via the anaplerotic entry of pyruvate or other substrates into the TCA cycle generating excess of intermediates that then exit the mitochondria to engage in various cytosolic pathways leading back to pyruvate. Pyruvate-isocitrate cycling generates an amplifying signal that enhances the Ca^{2+}-mediated triggering signal for insulin exocytosis

the liver has a central role in monitoring and regulating post-prandial (meaning "after a meal") glucose levels (Fig. 9.2, left). Insulin promotes hepatic triacylglycerol synthesis and storage of triacylglycerols in WAT upon feeding. Moreover, insulin also suppresses the release of stored lipids from adipose tissue. Ingested nutrients in intestinal endocrine cells stimulate the release of incretins, such as GLP1, that itself, together with the rise in blood glucose, stimulates β cells to deliver insulin.

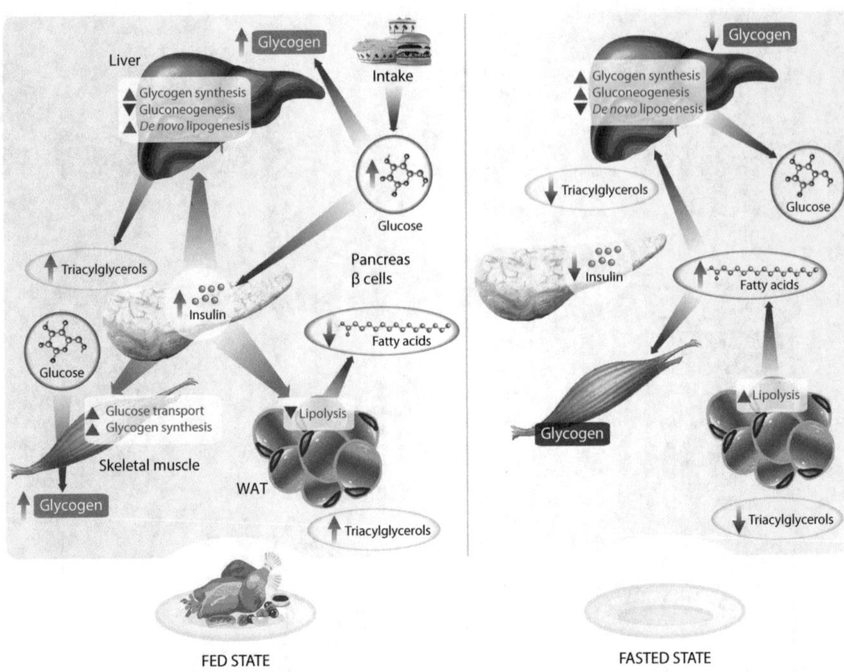

Fig. 9.2 Insulin action in health. In the fed state, dietary carbohydrates increase plasma glucose levels and promote insulin secretion from β cells. In skeletal muscle, insulin increases the transport of glucose and permits glucose entry and glycogen synthesis. In adipose tissue, insulin suppresses lipolysis and promotes *de novo* lipogenesis. In the liver, insulin stimulates glycogen synthesis and *de novo* lipogenesis and inhibits gluconeogenesis. In the fasted state, insulin secretion is decreased, which increases hepatic gluconeogenesis and promotes glycogenolysis. Under these conditions, hepatic lipid production diminishes while adipose lipolysis increases

The first phase of insulin secretion primarily prevents the liver from producing more glucose by stimulating glycogen synthesis and suppressing gluconeogenesis. The second phase, approximately 1–2 h after the meal, stimulates glucose uptake by insulin-sensitive tissues, such as skeletal muscle and adipose tissue. During fasting (Fig. 9.2, right), hepatic glycogenolysis decreases as hepatic glycogen stores deplete. Low insulin levels combined with elevated counter-regulatory hormones, such as glucagon, adrenaline and corticosteroids, promote hepatic glucose production via gluconeogenesis, so that blood glucose levels remain stable across a wide range of physiological conditions. Moreover, glucagon and adrenaline stimulate lipolysis in WAT and fatty acid oxidation in other tissues when nutrients are in limited supply. Although the hormonal regulation of glucose homeostasis is essential, also the CNS can sense and respond to acute changes in glucose and nutrient needs through innervation of the intestine, the liver, the pancreas, the portal vein and all other glucose-demanding tissues (Sect. 8.4)

9.2 Principles of Insulin Signaling

Insulin resistance is a major predictor for the development of T2D (Chap 10) and the metabolic syndrome (Chap 12). In order to develop new drugs to treat T2D and its cardiovascular complications, it is essential to understand the principles of insulin signaling. The major role of insulin signaling is the regulation of glucose, lipid and energy homeostasis, predominantly via action on skeletal muscle, liver and WAT. The final results of the insulin signaling are increased glucose uptake in muscle and fat tissue and inhibition of glucose synthesis in the liver. In adipose tissue, insulin also inhibits the release of FFAs.

Insulin acts via IRs which are located in the plasma membrane of insulin sensing cells. The IR is a tetrameric protein complex formed of each two α- and β-subunits and, together with IGF1R (Sect. 6.1), belongs to the receptor tyrosine kinase superfamily. When insulin binds to the α-subunit, the receptor undergoes a conformational change that activates the cytosolic kinase domain of the β-subunit and allows the recruitment of IRS proteins to its cytosolic component. The activity of IR is up-regulated by tyrosine phosphorylation, while the receptor is negatively regulated by protein tyrosine phosphatases. Moreover, SOCS1 and SOCS3, growth factor receptor-bound protein (GRB) 10 and ectonucleotide pyrophosphatase/phosphodiesterase 1 (ENPP1) inactivate the function of IR by blocking its interaction with IRS proteins or by modifying the receptor's kinase activity. SOCS proteins are up-regulated in insulin resistance (Sect. 9.4).

In particular during insulin resistance, IR is inactivated by ligand-stimulated internalization and degradation. IR has at least 11 intra-cellular substrates, such as IRSs 1–6, GRB2-associated binder 1 (GAB1), Cbl proto-oncogene, E3 ubiquitin protein ligase (CBL) and different isoforms of the Src homology 2 domain containing (SHC) protein. The interaction of phosphorylated IRS proteins with the regulatory subunit of PI3K results in the generation of the second messenger phosphatidylinositol-3,4,5-triphosphate (PIP3) activating AKT (Fig. 9.3). IR/IRS, PI3K and AKT form three important nodes in the insulin signal transduction cascade being responsible for most of the metabolic actions of insulin, such as glucose uptake, glucose synthesis and inhibition of gluconeogenesis. Main criteria for critical nodes are that they (i) have several isoforms that are involved in divergent signaling, (ii) are highly positively and negatively regulated, (iii) are essential for the biological actions of the ligand and (iv) mediate crosstalk with other signal transduction cascades. In this way, the three nodes explain the diversification and fine-tuning of insulin signaling both in health and disease.

IRS1 and IRS2 are ubiquitously expressed, while IRS3 is found primarily in adipocytes and the brain and IRS4 in embryonic tissues. IRS proteins are characterized by a 100 amino acids pleckstrin-homology (PH) domain close to their N-terminus and a central phosphotyrosine-binding (PTB) domain (Fig. 9.4). In addition, each IRS protein contains more than 20 tyrosine and serine phosphorylation sites. After IR phosphorylates IRS' tyrosine sites the protein interacts with Src-homology 2 (SH2) domain-containing adaptor proteins, such as the regulatory

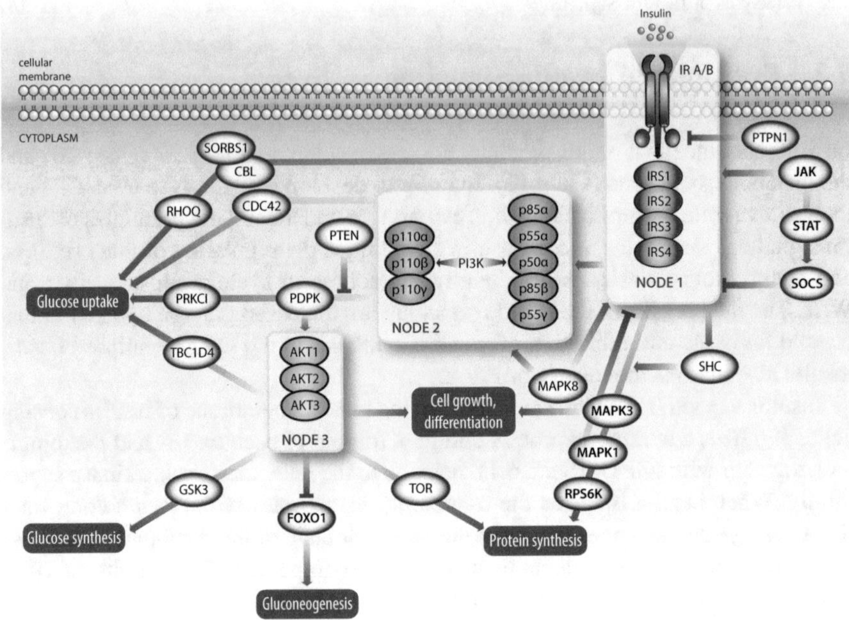

Fig. 9.3 Critical nodes in insulin signaling. Three important nodes of the insulin signaling network are IR/IRS, PI3K with its several regulatory and catalytic subunits and the three isoforms of AKT. Downstream of these nodes are the proteins TBC1D4, protein kinase C, iota (PRKC1), sorbin and SH3 domain containing 1 (SORBS1), cell-division cycle 42 (CDC42), glycogen synthase kinase 3 (GSK3), ribosomal protein S6 kinase (RPS6K) A1, 3-phosphoinositide dependent protein kinase (PDPK) 1 and 2, phosphatase and tensin homologue (PTEN), protein tyrosine phosphatase, non-receptor type 1 (PTPN1), ras homolog family member Q (RHOQ), CBL, JAK, FOXO1, MAPK1, MAPK3, MAPK8, TOR, SHC, SOCS and STAT

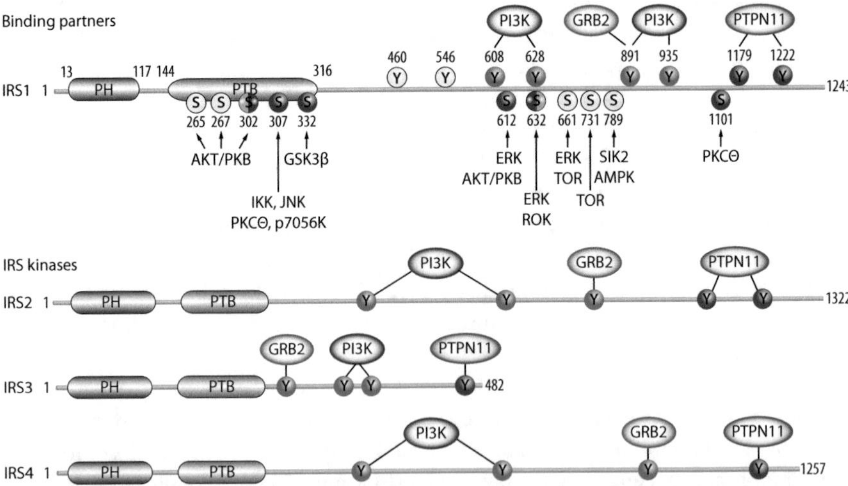

Fig. 9.4 IRS isoforms. The isoforms IRS1, IRS2, IRS3 and IRS4 share an N-terminal PH domain, a central PTB domain and several tyrosine (Y) and serine (S) phosphorylation sites. The kinases responsible for Y and S phosphorylations are shown; *blue circles* represent sites of positive regulation, *red circles* indicate sites of negative regulation, whereas function of the sites marked in *white* is not yet known. The proteins that bind or phosphorylate IRSs are listed

subunit of PI3K or GRB2. GRB2 then associates with the adaptor protein son of sevenless (SOS) to activate the MAPK pathway via MAPK1, MAPK3 and MAPK8. This links the action of insulin to the control of cell growth and differentiation. Like for IR, the signaling function of IRS proteins is also regulated by the action of tyrosine phosphatases, such as SH2-domain-containing tyrosine phosphatase 2 (SHP2). SHP2 dephosphorylates the IRS binding sites of PI3K and GRB2 and interrupts their respective signal transduction cascades. In response to insulin, FFAs and cytokines IRS proteins are phosphorylated at serine residues. Most of these serine phosphorylations negatively regulate IRS signaling, i.e. they represent a negative-feedback mechanism for the insulin signaling. Interestingly, serine phosphorylation of IRS1 strongly correlates with insulin resistance (Sect. 9.4). Moreover, reduced expression of IRS proteins or IR contributes to insulin resistance.

The kinase PI3K is formed by a regulatory and a catalytic subunit, each of which has several isoforms. The catalytic subunit is activated via the interaction of two SH2 domains in the regulatory subunit of IRS proteins. Inhibition of PI3K blocks most of insulin's actions on glucose transport, glycogen synthesis, lipid synthesis and adipocyte differentiation, i.e. the enzyme has a central role in the metabolic actions of insulin. The most important downstream target of PI3K is the serine/threonine kinase AKT, which phosphorylates a number of key proteins, such as GSK3, TBC1D4 and FOXO1. AKT has three isoforms, of which AKT2 seems to be most important in controlling metabolic functions. Phosphorylation of GSK3 decreases the kinases' inhibitory activity on GS leading to increased glycogen synthesis, but GSK3 has also a number of additional targets. Activated TBC1D4 stimulates small GTPases that are involved in cytoskeletal re-organization, which is required for the translocation of the glucose transporter GLUT4 to the plasma membrane, i.e. TBC1D4 controls glucose uptake. Finally, AKT controls the activity of FOXO1 (Sect. 9.3).

9.3 Central Role of FOXO Transcription Factors

The FOX transcription factor family contains over 100 members, some of which are crucial for the regulation of metabolism. The proteins FOXO1, FOXO3, FOXO4 and FOXO6 form a subclass of the family. FOXO1 is highly expressed in organs that control glucose homeostasis, such as in liver, skeletal muscle and adipose tissue, as well as in β cells.

FOXOs are activated by oxidative stress and ER stress via MAPK8 and are negatively regulated by the insulin signaling pathway via PI3K and AKT (Fig. 9.5a). In *C. elegans* and other model organisms the IR-PI3K-AKT-FOXO signaling axis shows central functions in aging: optimal FOXO signaling ensures longer lifespan, while the de-regulation of this pathway contributes to the age-related diseases cancer and T2D (Sect. 6.1). This provides FOXO with a central role at the interconnection of aging and disease and suggests that the main function of FOXOs is to maintain homeostasis in response to environmental stress, such as an increased oxi-

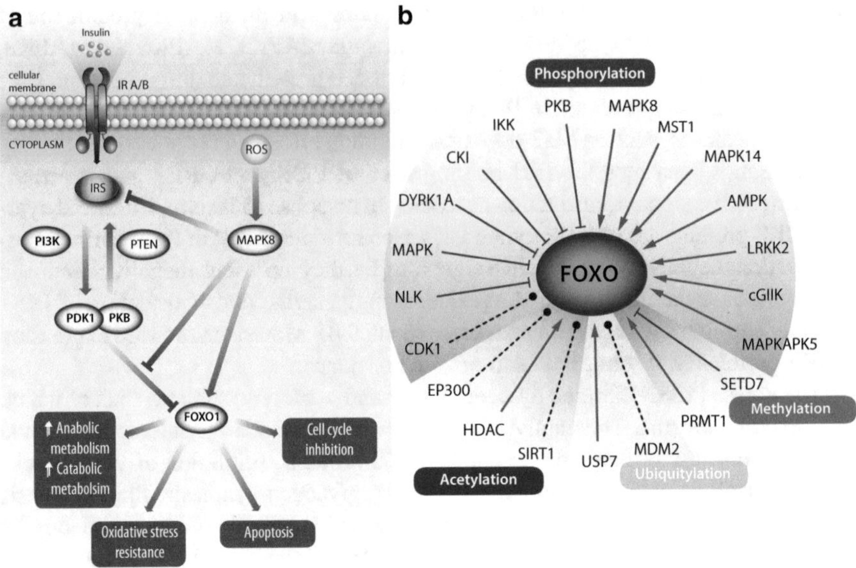

Fig. 9.5 Control and outcome of FOXO activation. (**a**). Antagonistic control of FOXO through PI3K-AKT signaling and MAPK8 signaling. Insulin and growth factors signal converge in their signaling at PI3K and activate PDPK1 and AKT. This is counteracted by PTEN. Active AKT can inhibit the transcriptional activity of FOXOs through phosphorylation at three conserved residues, resulting in cytoplasmic retention of FOXO. Oxidative stress induces the nuclear translocation of FOXOs via MAPK8 and thereby activates FOXO target genes. MAPK8 can inhibit the action of insulin at multiple levels and thereby counteracts the inhibition of FOXO. The activation of FOXO inhibits the cell cycle, increases oxidative stress resistance, induces apoptosis and shifts cellular metabolism away from anabolic pathways towards catabolic metabolism. (**b**). In addition to AKT and MAPK8, FOXOs are also regulated via post-translational modifications, such as phosphorylation, acetylation, methylation and ubiquitylation, by various signal transduction pathways

dative stress, starvation or overnutrition. In the liver, FOXO1 is activated during starvation or low glucose levels. Under these conditions the transcription factor changes metabolism so that it ensures glucose homeostasis, i.e. it initiates the breakdown of glycogen and gluconeogenesis. Moderate insulin resistance leads to reduced insulin signaling resulting via FOXO-stimulated gluconeogenesis in hyperglycemia. In addition, in severe insulin resistance, high levels of FOXO-triggered lipid oxidation lead to ketoacidosis. FOXO1 also affects the fasting response via actions in the CNS, such as in AGRP neurons (Sect. 8.4). Therefore, a balanced regulation of FOXOs, especially of FOXO1, via the IR-IRS-PI3K-AKT axis, is essential for normal transcriptional control during the metabolic response.

FOXOs are regulated by a large number of signal transduction pathways and post-translational modifications (Fig. 9.5b). This is similar to other intensively studied transcription factors, such as p53, and suggests that combinatorial codes of post-translational modification regulate the function of key transcription factors. For example, the meaning of FOXO acetylation may be to switch FOXO-induced gene expression from an apoptotic to a pro-survival response. Such a code would be analogous to the histone code (Box 5.1). FOXOs interact with SIRT1 during oxidative stress, but only when active PI3K signaling also promotes uptake of SIRT1 into the nucleus. Moreover, the energy sensor AMPK (Sect. 6.5) phosphorylates FOXOs, i.e. AMPK directs the transcriptional action of FOXOs, so that it activates alternative energy sources and stress resistance, as it was already observed under conditions of calorie restriction. Interestingly, AMPK activates FOXOs without influencing its subcellular localization, but it affects the shuttling of FOXO co-regulators. For example, AMPKs retains HDAC4, HDAC5 and HDAC7 in the cytosol and enhances the interaction of the HAT CREB-binding protein (CBP) with FOXOs in the nucleus, i.e. the acetylation of FOXOs is maximized. Taken together, FOXO transcription factors function as scaffolds that are post-transcriptionally modified by a number of common signal transduction cascades. This means that the activity of FOXOs depends a lot on the cellular context and the complete set of signals that a cell type or tissue is exposed to.

Under conditions of tissue homeostasis FOXOs are inactive and located in the cytoplasm (Fig. 9.6a). The signal transduction of both insulin and growth factors converge at PI3K, the activation of which leads to an increased level of PIP3. This second messenger then activates PDPK1, which in turn stimulates AKT. Active AKT translocates to the nucleus and phosphorylates FOXOs at three conserved residues, resulting in increased binding of the transcription factors to tyrosine 3-monooxygenase/tryptophan 5-monooxygenase activation protein (YWHA, also called 14-3-3). YWHA proteins bind more than 200 functionally diverse signaling proteins, such as transcription factors, kinases and transmembrane receptors, when they are phosphorylated at serine or threonine residues and retain them in inactive form in the cytoplasm. In contrast, under cellular stress, in particular at high ROS levels, FOXOs translocate into the nucleus and activate their target genes (Fig. 9.6a). MAPK8 counteracts the activity of insulin at multiple levels, such as by decreasing IRS activity (Sect. 9.2) and by inducing the release of FOXOs from YWHA proteins.

All FOXOs recognize the same octameric binding sequence, i.e. they can replace each other. The FOXO binding site occurs quite often in the genome suggesting that the transcription factors can act as pioneer factors, i.e. as factors that make the first contact with condensed chromatin, open it up via the recruitment of HATs, such as CBP, and help other transcription factors to bind at these regions. In this way, FOXOs functionally interact with a number of transcription factors. These include the nuclear receptors RAR and PPAR and the cell growth-regulating transcription factors p53 and MYC. p53 and FOXOs have a comparable functional profile, but MYC and FOXOs act as reciprocal antagonists, i.e. both potent transcription factors keep each other under control. Thus, FOXOs contribute to cellular homeostasis by integrating different signal transduction cascades that are sensitive to environmental changes, such as starvation, oxidative stress and growth factor deprivation. FOXO expression does not predict a certain cellular function, but it rather enables the amplification of the cellular potential, i.e. the transcription factor provides an integration point for several cellular inputs and induces a distinct gene expression program.

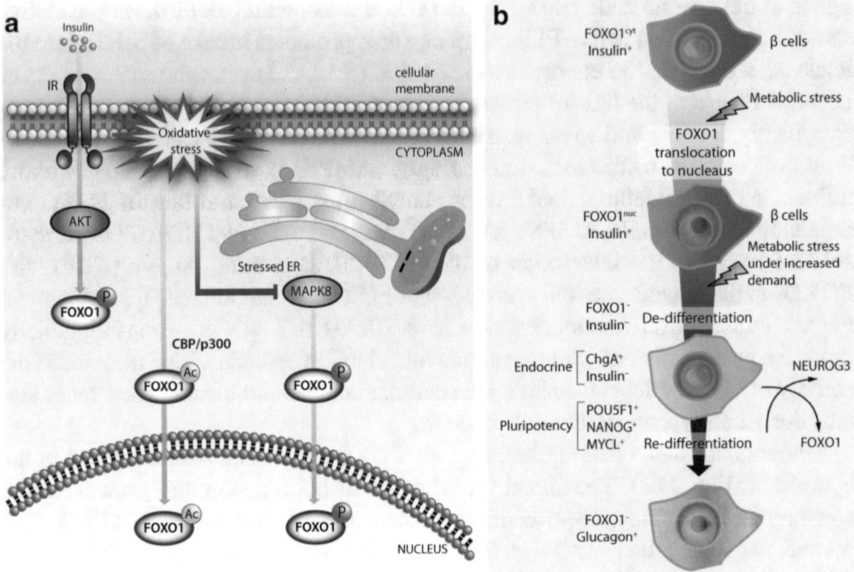

Fig. 9.6 FOXO1 signaling is affected by metabolic and oxidative stress. (**a**). Continuous insulin stimulation activates AKT, which in turn phosphorylates serine and threonine sites of FOXO1, so that the transcription factor is retained to the cytoplasm. However, under oxidative stress conditions FOXO1 is acetylated by CBP and locates preferentially in the nucleus. The acetylation prevents the ubiquitination of FOXO1, i.e. the protein is stabilized and retained in the nucleus. In addition, oxidative stress also activates MAPK8, which phosphorylates FOXO1 at serine and threonine sites distinct from those phosphorylated by AKT and further enhances nuclear retention of FOXO1. Furthermore, also ER stress can induce nuclear retention of FOXO1 via MAPK8 activation. (**b**). Under severe hyperglycemic conditions, β cells no longer express FOXO1, insulin, PDX1 and MAFA but the pluripotency markers POU5F1, NANOG and MCYL, indicating de-differentiation into progenitor-like cells. Some of these cells start to express glucagon suggesting they re-differentiate into α cells

In mild hyperglycemia, i.e. at metabolic stress, FOXO1 translocates into the nucleus of β cells. Under condition of additional stress, such as severe hyperglycemia, FOXO1 is down-regulated and the cells loose the expression of insulin and the transcription factors PDX1 and v-maf avian musculoaponeurotic fibrosarcoma oncogene homolog A (MAFA) (Fig. 9.6b). Then β cells change their transcriptional profile and instead express the endocrine progenitor markers chromogranin A and neurogenin 3 and the pluripotency markers POU class 5 homeobox 1 (POU5F1), Nanog homeobox (NANOG) and v-myc avian myelocytomatosis viral oncogene lung carcinoma derived homolog (MYCL). This is one mechanism of β cell failure (Sect. 9.5), in which the cells de-differentiate into progenitor-like cells and re-differentiate into glucagon-producing α cells.

9.4 Insulin Resistance in Skeletal Muscle and Liver

Insulin resistance is a condition, in which normal concentrations of insulin produce a subnormal biological response in insulin target tissues. The β cells compensate the subnormal response by increasing the production of insulin. As long as the hyperinsulinemia is adequate to overcome the insulin resistance, glucose tolerance remains relatively normal. In patients destined to develop T2D, the β cell compensatory response fails (Sect. 9.5) and insulin insufficiency develops leading to impaired glucose tolerance and eventually T2D. Impaired insulin sensitivity causes impaired insulin-stimulated glucose uptake into skeletal muscle, impaired insulin-mediated inhibition of hepatic glucose production in liver and a reduced ability of insulin to inhibit lipolysis in WAT.

There are three main mechanisms to explain insulin resistance, in particular in muscle cells (Fig. 9.7) and liver cells (Fig. 9.8). These are (i) an ectopic (meaning "unusual") lipid accumulation in muscle and liver, (ii) a chronic inflammatory response of the tissues (Sect. 7.2) and ER stress mediated via the unfolding protein response (Sect. 7.5). All three mechanisms are originally derived from an evolutionary advantage for the organism in adapting to a changing environment. The evolutionary oldest pathway, the unfolded protein response, was designed to integrate metabolic signals, such as metabolic stress through lipid overload, and to adapt accordingly. The inflammatory response also represents an evolutionary old pathway that is highly interconnected with the unfolded protein response, in order to provide a coordinated response to various environmental stimuli, such as nutrient scarcity. This requires the dampening of the insulin response, in order to allow a metabolic shift from glucose to lipid oxidation.

Therefore, humans that undergo long-term fasting become insulin resistant in one or several tissues, such as muscle, liver or WAT, in order to preserve blood glucose for the brain. This insulin resistance leads to increased fatty acid concentrations in the circulation as well as in skeletal muscle and liver. The elevated lipid levels in these tissues also cause a parallel increase in lipids with signaling function, such as diacylglycerol (DAG) and ceramides. This then enhances the insulin resistance of the tissues and ensures the preservation of glucose for the CNS. However, this natural mechanism for survival became pathogenic in the modern times, where often the level of energy intake exceeds the level of energy expenditure, i.e. in conditions of misbalanced energy homeostasis (Sect. 8.3).

The lipid content in muscle cells reflects a net balance between fatty acid uptake and their oxidation in mitochondria. Thus, acquired mitochondrial dysfunction is an important predisposing factor for ectopic lipid accumulation and insulin resistance in the elderly. LPL (Sects. 3.4 and 7.3) is a key enzyme for the hydrolysis of circulating triacylglycerols (for example within VLDLs) that permits the uptake of fatty acids in muscle and liver through a complex of fatty acid transport proteins of the SLC27A family with the scavenger receptor CD36 (Figs. 9.7 and 9.8). Upon entry into the cell, fatty acids are rapidly esterified to acyl-CoAs. These are successively transferred to a glycerol backbone to form mono-, di-, and triacylglycerols or

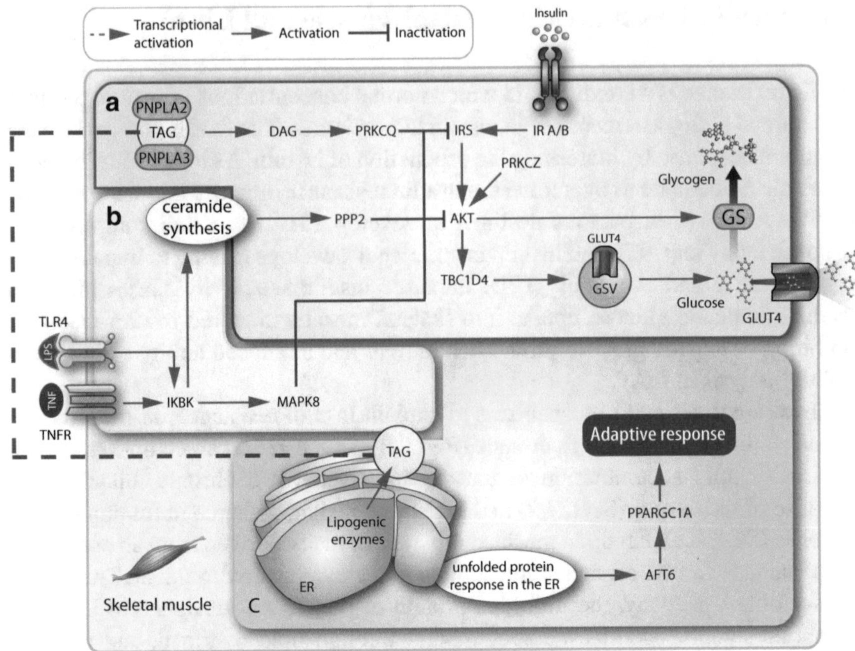

Fig. 9.7 Pathways involved in muscle insulin resistance. The insulin-IR-IRS-PI3K-AKT signaling axis promotes via TBC1D4 the translocation of GLUT4-containing storage vesicles (GSVs) to the plasma membrane permitting the entry of glucose into the cell and also stimulates glycogen synthesis via GS. This central signal transduction cascade is connected to a number of other pathways. (**a**). *Green shaded*: DAG-mediated activation of PRKCQ and the subsequent inhibition of IRS, ceramide-mediated increases in the AKT inhibitor PPP2 and increased sequestration of AKT by PRKCZ. Impaired AKT2 activation limits translocation of GSVs to the plasma membrane, resulting in impaired glucose uptake. It also decreases insulin-mediated glycogen synthesis. (**b**). *Yellow shaded*: Inflammatory pathways, such as the activation of IKBK affecting ceramide synthesis and the activation of MAPK8 inhibiting IRS via phosphorylation. (**c**). *Pink shaded*: The unfolded protein response in the ER leading to activation of ATF6 and a PPARGC1A-mediated response. Key lipogenic enzymes in the ER membranes stimulate lipid droplet formation

esterify with sphingosine to form ceramides. This means that the levels of the second messengers DAG and ceramide rise in parallel to the increased lipid load of the cells. Intracellular lipid droplets possess a mantle of enzymes, such as the lipases patatin-like phospholipase domain containing (PNPLA) 2 and PNPLA3, that regulate the entry and exit of lipid molecules and catalyze their lysis, for example from triacylglycerols to DAG. Thus, these enzymes are essential both for the access of the energy stored in triacylglycerols and the generation of lipid mediators of insulin resistance. Interestingly, PNPLA2 is an AMPK target.

Both in muscle and in liver DAG activates members of the protein kinase C family, such as PRKCQ, that impair insulin signaling via inhibition of IRS1 and IRS2 resulting in a decreased glucose uptake via GLUT4 (Figs. 9.7 and 9.8). Ceramides dampen insulin signaling (i) by the activation of protein phosphatase 2 (PPP2)

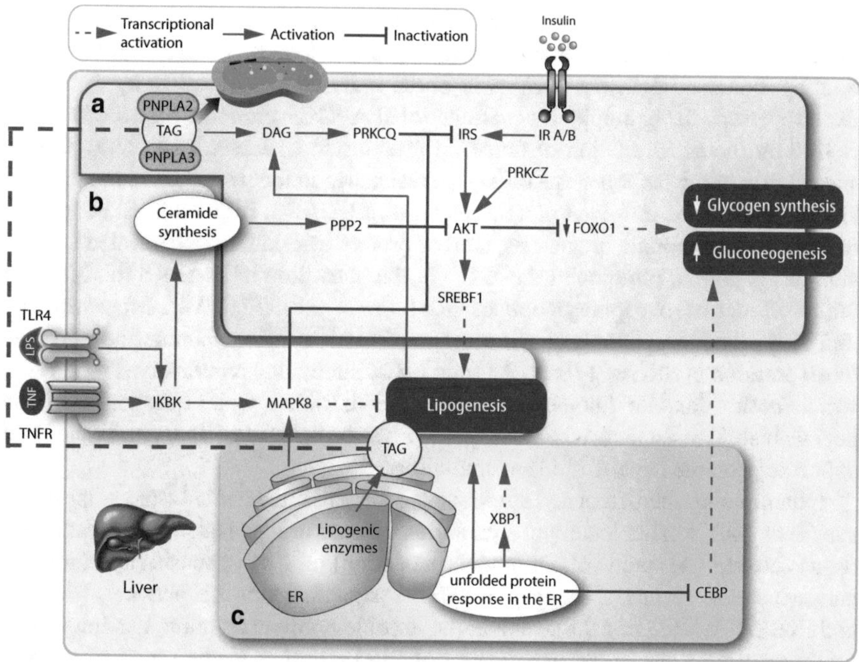

Fig. 9.8 Pathways involved in hepatic insulin resistance. The insulin-IR-IRS-PI3K-AKT signaling axis promotes glycogen synthesis (not shown), suppresses gluconeogenesis and activates *de novo* lipogenesis. This central signal transduction cascade is connected to a number of other pathways. (**a**). *Green shaded*: DAG-mediated activation of PRKCQ and subsequent inhibition of IRS, ceramide-mediated increases in PPP2 and increased sequestration of AKT by PRKCZ. Impaired AKT2 activation (i) limits the inactivation of FOXO1 leading to increased expression of key enzymes of gluconeogenesis and (ii) decreases insulin-mediated glycogen synthesis (not shown). (**b**). *Yellow shaded*: Inflammatory pathways, such as the activation of IKBK affecting ceramide synthesis and the activation of MAPK8 impairing lipogenesis. (**c**). *Pink shaded*: The unfolded protein response in the ER leads to increased lipogenesis via XBP1 and also increased gluconeogenesis via CEBPs. The ER membrane also contains key lipogenic enzymes stimulating lipid droplet formation. Proteins that regulate the lipid release from these droplets, such as PNPLA2 and PNPLA3, can modulate the concentration of key lipid intermediates in discrete cell compartments

dephosphorylating AKT and (ii) via PKCζ that binds AKT and prevents its activation. When in the liver the rates of DAG synthesis from fatty acid re-esterification and *de novo* lipogenesis exceed the rates of lipid oxidation in the mitochondria, i.e. when there is an increase of DAG levels, PRKCE is activated, IR tyrosine kinase activity is inhibited, GSK3 is hyperphosphorylated and glycogen synthesis is decreased. Furthermore, this leads to increased translocation of FOXO to the nucleus promoting elevated expression of gluconeogenic enzymes, such as PCK2 and G6PC (Fig. 9.8).

Ectopic accumulation of lipids in the liver is termed non-alcoholic fatty liver disease (NAFLD). Many individuals with this disease also have increased visceral adiposity (Sect. 8.1), but hepatic insulin resistance is primarily related to intra-hepatic lipid content, not to the visceral fat mass. Lipid overload in the liver can be caused both by increased delivery, such as in obese individuals, as well as by decreased export caused, for example, by malfunctional APOC3 proteins. Stress of the ER caused by the accumulation of unfolded proteins in its lumen (Sect. 7.5) plays a special role in the pathogenesis of insulin resistance in the liver. Activation of three key proteins of the unfolded protein response, EIF2AK3, ERN1 and ATF6, results in increased membrane biogenesis, stop of protein translation and elevated expression of chaperone proteins in the ER. Via the activation of MAPK8 this leads to inhibitory serine phosphorylation of IRS1 (Sect. 9.2) (Fig. 9.8). Moreover, the unfolded protein response results in an expansion of the ER membrane and increases the expression of SREBF1 (Sect. 3.1) and MLX interacting protein-like (MLXIPL), which both stimulate lipogenesis. Thus, the unfolded protein response causes hepatic insulin resistance, when it is able to alter the balance of lipogenesis and lipid export to promote hepatic lipid accumulation.

Inflammatory mediators and adipokines, such as TNF, secreted from adipose tissue (Sect. 8.6), can act locally in a paracrine manner or they leak out of the adipose tissue causing a systemic effect (endocrine action) on insulin sensitivity in muscle and liver cells. Via the TNF receptor (TNFR) signaling axis this activates MAPK8 and IKBK (Figs. 9.7 and 9.8). Like in the unfolded protein reaction, this inflammatory response results in the inactivation of IRS1 and thus leading to insulin resistance. MAPK8 and IKBK also phosphorylate and activate transcription factors, such as AP-1 and NF-κB, leading to increased expression of inflammatory mediators. In parallel, activation of TLR4 by PAMPs, such as SFAs, in skeletal muscle and liver cells further enhances the NF-κB pathway.

9.5 β Cell Failure

The failure of β cells during the progression to T2D involves their chronic exposure to glucose and lipids, also known as "glucotoxicity" and "lipotoxicity", or in combination "glucolipotoxicity". The chronic glucose exposure increases glucose metabolism in β cells, leading to the formation of citrate, which acts as a signal for the formation of malonyl-CoA in the cytosol (Fig. 9.9). Malonyl-CoA inhibits the key fatty acid transporter in mitochondria, carnitine palmitoyltransferase 1A (CPT1A), and blocks in this way fatty acid β-oxidation. This causes accumulation of SFA-CoAs in β cells. The high demand for insulin secretion during hyperglycemia creates significant metabolic stress to the ER of β cells and results in the overproduction of ROS in their mitochondria.

This oxidative stress is a central element of glucotoxicity. When intracellular glucose concentrations exceed the glycolytic capacity of β cells, some of the molecules are converted to enediol intermediates, leading to superoxide formation. Since β cells contain only low levels of anti-oxidant enzymes, such as catalase, glutathione peroxidase and SOD2, they are very susceptible to superoxide damage.

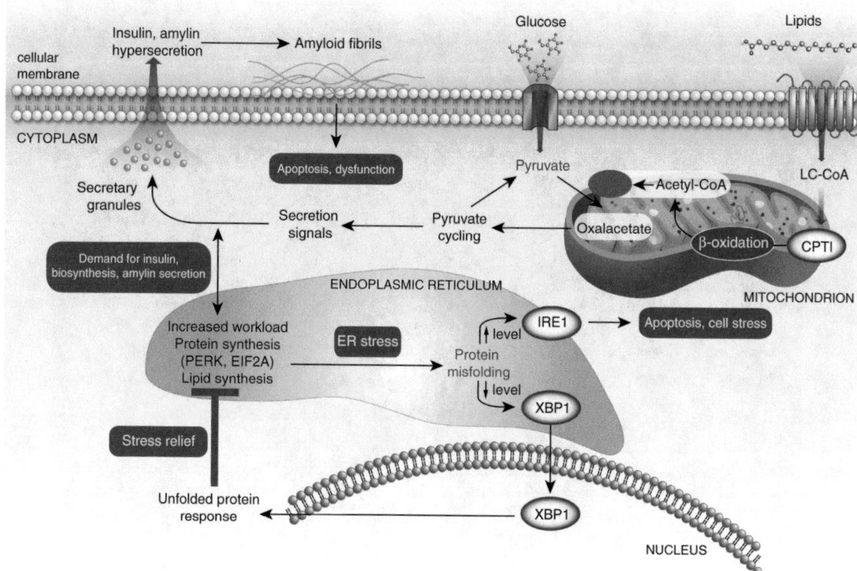

Fig. 9.9 Mechanisms of β cell failure. Overnutrition and/or increased lipid supply induces in mitochondria of β cells enzymes of β oxidation, such as CPT1A, resulting in increased acetyl-CoA levels, allosteric activation of PC and constitutive up-regulation of pyruvate cycling. This leads to increased basal secretion of insulin and a loss of the glucose-stimulated increment in the flux of pyruvate cycling, i.e. blunting of glucose-stimulated insulin secretion. The elevated demand for synthesis of insulin in the ER increases the stress to this organelle, thus resulting in elevated rates of protein misfolding. The protein unfolding response is initially able to balance this ER stress, but over time this becomes less effective, and the deleterious effects of ER stress leads to cell death. Insulin hypersecretion is accompanied by amylin secretion forming amyloid fibrils that accumulate at the surface of β cells and induce dysfunction and apoptotic death to the cells

Moreover, β cells have many mitochondria and consume more oxygen than most other cell types. Therefore, at high glucose conditions, such as directly after a meal, the accelerated mitochondrial function enhances the oxygen consumption and causes hypoxia. This parallels with the increased expression of hypoxia-inducible genes. In addition, T2D patients have an expanded ER in their β cells indicating increased stress to the organelle. The hyperinsulinemia in response to chronic hyperglycemia disrupts ER homeostasis in β cells due to the exceeded capacity for proinsulin biosynthesis. This leads to accumulation of misfolded proteins and induction of the unfolded protein response (Sect. 7.5), which enhances the oxidative stress and eventually results in β cell dysfunction.

Patients with a long clinical history of T2D (Chap 10) commonly have a decreased β cell number and function, respectively, which is often referred to as "β cell exhaustion". Compared with weight-matched healthy individuals, obese T2D patients have a 63% reduction of β cell mass, while lean T2D patients only show a 41% loss. This suggests that β cell dysfunction has a primary role in the pathogen-

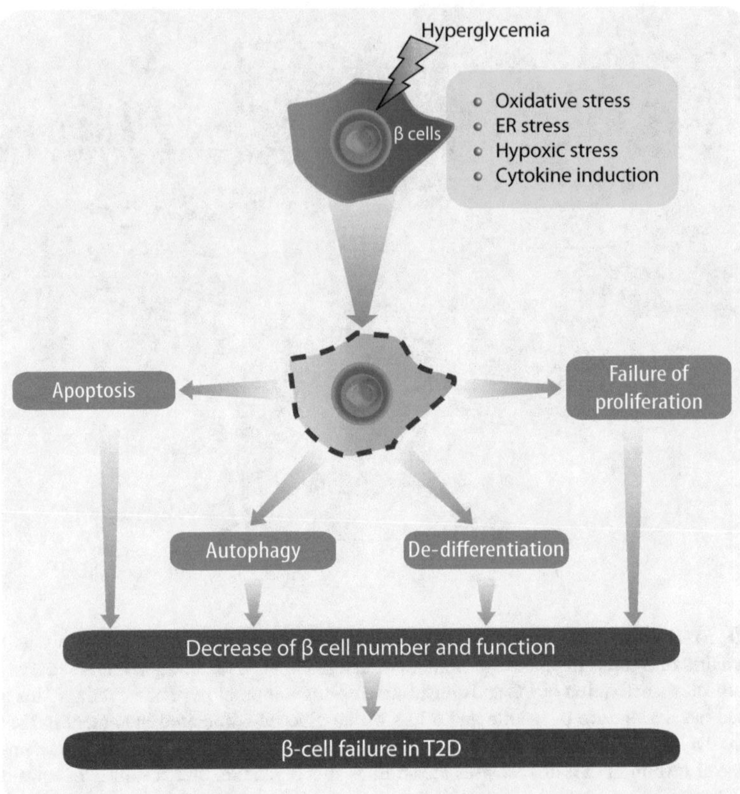

Fig. 9.10 Principles of β cell failure in T2D. Prolonged hyperglycemia results in oxidative, ER and hypoxic stress and in increased exposure of β cells with cytokines. Therefore, β cells may cease their proliferation, de-differentiate or undergo uncontrolled autophagy or apoptosis. All these processes reduce the number of β cells and their function. This leads dysfunction and depletion of β cells and to progress of T2D

esis of T2D. In response to ER stress, hypoxic stress and exposure to pro-inflammatory cytokines (Sect. 9.3), β cells fail to proliferate or undergo apoptosis or uncontrolled autophagy (Fig. 9.10). Moreover, the β cells can de-differentiate or trans-differentiate into other pancreatic cell types, such as α cells (Sect. 9.3). β cells proliferate via the replication of pre-existing β cells and the differentiation of progenitor cells, referred to as neogenesis. In the pancreas of adults, the dominant mechanism for increasing β cell numbers is replication rather than neogenesis. The mass of β cells is controlled by the balance between the rate of proliferation and the rate of apoptosis. The Fas cell surface death receptor (FAS) pathway, a central apoptosis regulatory mechanism, is up-regulated in patients with poorly controlled T2D. Thus, both increased apoptosis or decreased proliferation can reduce the β cell mass of T2D patients. High levels of saturated FFAs, i.e. lipotoxicity, also induce β cell apoptosis. Interestingly, in patients who are genetically predisposed to T2D, but not in healthy individuals, a sustained increase in plasma FFA levels causes β cell dysfunction.

In β cells, FOXO1 proteins are constitutively localized to the cytoplasm, which is in contrast to hepatocytes, muscle cells and adipocytes (Sect. 9.3). This is due to the continuous endogenous production of insulin, which activates AKT and keeps FOXO1 in the cytoplasm. However, in response to oxidative stress, FOXO1 is acetylated by CBP/p300-interacting transactivator 1 (CITED1) and translocates to the nucleus. The acetylation of FOXO1 also prevents ubiquitination of the protein and thus enhances its nuclear retention (Sect. 9.3). In addition, oxidative stress and ER stress activate MAPK8, which phosphorylates FOXO1 and further keeps the protein in the nucleus. Since FOXO1 regulates the expression of anti-oxidant enzymes, such as catalase and SOD2, this should protect β cells against various stressors.

Future View

The insulin signaling axis via IR, IRS, PI3K, AKT and FOXO is a paradigm for the complexity and flexibility of a molecular signaling pathway. Although this signaling cascade is better understood than many other pathways, in future we need to get significantly more insight, in order to monitor the molecular response of metabolic organs to metabolic and inflammatory stress in health and disease. A critical issue will be the understanding of the post-translational modification code of key regulatory proteins, such as FOXOs. Furthermore, we need to better understand, (i) the mechanisms that drive and regulate the lipid accumulation in skeletal muscle and liver, (ii) the genes involved in lipid uptake, transport and export, (iii) how diet regulates lipid accumulation causing or protecting against insulin resistance and (iv) the mechanisms of β cell failure and the genes involved in this process.

Key Concepts

- Glucose homeostasis controls glucose production and glucose use even at physiological challenges, such as food ingestion, fasting and intense physical exercise, and maintains the blood glucose level within the range of 4–6 mM.
- The first phase of insulin secretion primarily prevents the liver from producing more glucose by stimulating glycogen synthesis and suppressing gluconeogenesis. The second phase, approximately 1–2 h after the meal, stimulates glucose uptake by insulin-sensitive tissues, such as skeletal muscle and adipose tissue.
- Insulin resistance is a major predictor for the development of T2D and the metabolic syndrome.
- IR/IRS, PI3K and AKT form three important nodes in the insulin signal transduction cascade being responsible for most of the metabolic actions of insulin, such as glucose uptake, glucose synthesis and inhibition of gluconeogenesis.
- The most important downstream target of PI3K is the serine/threonine kinase AKT, which phosphorylates a number of key proteins, such as GSK3, TBC1D4 and FOXO1.
- FOXOs have a special role at the interconnection of aging and disease. Their main function is to maintain homeostasis in response to environmental stress, such as an increased oxidative status, starvation or overnutrition.

- FOXO transcription factors function as scaffolds that are post-transcriptionally modified by a number of common signal transduction cascades.
- The three main mechanisms to explain insulin resistance are (i) an ectopic lipid accumulation in muscle and liver, (ii) a chronic inflammatory response of the tissues and ER stress mediated via the unfolding protein response.
- The levels of the second messengers DAG and ceramide rise in parallel to the increased lipid load of the cells.
- The unfolded protein response causes hepatic insulin resistance when it is able to alter the balance of lipogenesis and lipid export to promote hepatic lipid accumulation.
- The failure of β cells during the progression to T2D involves the chronic exposure of the cells to glucose and lipids (glucotoxicity and lipotoxicity).
- The high demand for insulin secretion during hyperglycemia creates significant metabolic stress to the ER of β cells and results in the overproduction of ROS in mitochondria.
- The hyperinsulinemia in response to chronic hyperglycemia disrupts ER homeostasis in β cells due to the exceeded capacity for proinsulin biosynthesis leading to the induction of the unfolded protein response.
- In response to ER stress, hypoxic stress and exposure to pro-inflammatory cytokines, β cells fail to proliferate or undergo apoptosis or uncontrolled autophagy.

Additional Reading

Eijkelenboom A, Burgering BM (2013) FOXOs: signalling integrators for homeostasis maintenance. Nat Rev Mol Cell Biol 14:83–97

Johnson AM, Olefsky JM (2013) The origins and drivers of insulin resistance. Cell 152:673–684

Kitamura T (2013) The role of FOXO1 in beta-cell failure and type 2 diabetes mellitus. Nat Rev Endocrinol 9:615–623

Muoio DM, Newgard CB (2008) Mechanisms of disease: molecular and metabolic mechanisms of insulin resistance and beta-cell failure in type 2 diabetes. Nat Rev Mol Cell Biol 9:193–205

Odegaard JI, Chawla A (2013) Pleiotropic actions of insulin resistance and inflammation in metabolic homeostasis. Science 339:172–177

Perry RJ, Samuel VT, Petersen KF, Shulman GI (2014) The role of hepatic lipids in hepatic insulin resistance and type 2 diabetes. Nature 510:84–91

Samuel VT, Shulman GI (2012) Mechanisms for insulin resistance: common threads and missing links. Cell 148:852–871

Shulman GI (2014) Ectopic fat in insulin resistance, dyslipidemia, and cardiometabolic disease. N Engl J Med 371:1131–1141

Taniguchi CM, Emanuelli B, Kahn CR (2006) Critical nodes in signalling pathways: insights into insulin action. Nat Rev Mol Cell Biol 7:85–96

Chapter 10
Diabetes

Abstract Diabetes is a disease of dys-regulation of glucose and lipid homeostasis that does not only affect the insulin production in the β cells but also the metabolism in organs such as liver, muscle and fat. Worldwide, the prevalence of T2D is rapidly increasing, which, when not properly treated, ultimately leads to reduced life expectancy due to microvascular (retinopathy, nephropathy and neuropathy) and macrovascular (heart disease and stroke (Chap. 11) complications. Like in obesity (Chap. 8), both genetic and environmental factors contribute to the development of diabetes. For example, persons at high risk for developing T2D should benefit from lifestyle changes involving healthy diet, moderate weight loss and increased physical activity. Despite large GWAS screening for risk genes, at present less than 10 % of the inheritance of T2D is understood. Therefore, in addition, epigenome-wide changes, both pre-natal as well as in adult life, are intensively investigated.

In this chapter, we will describe the different forms of diabetes, their diagnosis and the worldwide prevalence of the disease. We will discuss the dys-regulation of glucose homeostasis in T2D. In this context, we will present the genetic and physiologic basis of the disease and again we will highlight chronic inflammation as the core of the disease, this time affecting islets of the pancreas. We will realize that the present understanding of T2D risk genes is insufficient and that most likely epigenetics plays an important role in the disease, as examplified through the thrifty gene hypothesis.

Keywords T1D • T2D • OGTT • Insulin • β cells • Liver • Skeletal muscle • Adipose tissue • Inflammation • MODY • GWAS • Epigenetic programming • Thrifty gene hypothesis

10.1 Definition of Diabetes

Diabetes is a condition of chronically elevated plasma glucose levels, referred to as hyperglycemia, that eventually causes toxicity to blood vessels (Box 10.1). There are two major forms of diabetes, type 1 (T1D) and T2D. T1D results from an autoimmune destruction of insulin-producing β cells in the pancreas. As a result, the body can no longer produce insulin and the respective patients need insulin injections every day for the rest of their life, in order to control the levels of glucose in

C. Carlberg et al., *Nutrigenomics*, DOI 10.1007/978-3-319-30415-1_10

Box 10.1 Diabetes complications

People with diabetes (both T1D and T2D) are at risk of developing a number of troubling, disabling and life-threatening health problems. Chronically high blood glucose levels can lead to serious diseases affecting the blood vessels of brain, heart, eyes, kidneys and peripheral nerves. In almost all high-income countries, diabetes is a leading cause of CVD, blindness, kidney failure and lower-limb amputations.

CVDs: This is the most common cause of disability and death among people with diabetes. CVDs that accompany diabetes include cerebral stroke, myocardial ischemia, congestive heart failure and peripheral artery disease (Chap. 11).

Kidney insufficiency: Nephropathy finally resulting in kidney failure is caused by damage to small blood vessels, through which the kidneys function less efficiently or even fail. Diabetes is one of the leading causes of chronic kidney disease.

Eye disease: In retinopathy the network of blood vessels, which supply the retina becomes blocked and damaged leading to increasingly loss of vision, finally to complete blindness.

Nerve damage: In neuropathy nerves throughout the body are damaged, which can lead to problems with digestion and micturition, erectile dysfunction and a number of other dysfunctions. The most commonly affected areas are the extremities, particularly the lower legs and feet (peripheral neuropathy) leading to pain, paresthesia, and loss of feeling. The latter is particularly dangerous because even small injuries will be unnoticed, leading to ulcerations, superinfections and finally to major amputations (diabetic foot syndrome).

their blood. Without insulin, a person with T1D will die. This type of diabetes has a sudden onset and usually affects children of 10 years or older, i.e. in an age when their immune system has reached full potency. The number of people who develop T1D is increasing, which may be due to changes in environmental risk factors, prenatal events, diet early in life or viral infections.

T2D is the most common type of diabetes, representing more than 90 % of all diabetes cases. It usually occurs in adults, but is increasingly seen in children and adolescents. In initial stages of T2D, β cells are still able to produce insulin, but either the amounts are insufficient or the body is unable to respond to its effects (known as insulin resistance, Sect. 9.4), both leading to elevated glucose levels in the blood. T2D often remains unnoticed and undiagnosed for years, i.e. the respective persons are unaware of the already smouldering long-term damage being caused by their disease. In contrast to people with T1D, the majority of T2D patients usually do not require daily doses of insulin to survive. Many T2D patients are able to manage their hyperglycemia through a healthy diet and increased physical activity or by oral medication for a rather long time (Sect. 10.2). However, when they reach

Box 10.2 Measuring Glucose Homeostasis
Methods to measure glucose homeostasis are:

Oral glucose tolerance test (OGTT): This test measures the speed with which glucose is cleared from the blood (Fig. 10.1). A defined amount of glucose is administered orally and blood is sampled at multiple time points. The test provides a general assessment of glucose homeostasis and is easy to perform.

Insulin sensitivity: In humans, overnight fasting establishes a steady-state metabolic condition, from which glucose and insulin levels are measured, for example during an OGTT, and incorporated into mathematical models that allow estimating insulin sensitivity. An insulin clamp allows a more accurate measurement, where a continuous intravenous infusion of insulin is provided together with glucose, in order to maintain basal glucose levels. Whole-body insulin sensitivity is then reflected by the overall amount of exogenous glucose infusion.

Insulin secretion: Insulin secretion can be tested by an OGTT, where plasma insulin levels are measured at multiple time points. Alternatively, a hyperglycemic clamp can be used, where a continuous intravenous infusion of glucose is provided in combination with measurements at multiple time points.

a stage, in which they are unable to regulate their blood glucose levels, they need external insulin substitution. Women, who during pregnancy may develop a resistance to insulin (mostly around the 24th week due to hormones produced by the placenta) and subsequent develop high blood glucose levels, have gestational diabetes. Uncontrolled gestational diabetes can have serious consequences for both the mother and her baby and increases the risk of the child to develop T2D later in life.

An oral glucose tolerance test (OGTT) with measurements of glucose and insulin at defined times (for example, 0, 30, 60 and 120 min) after oral uptake of a defined amount of glucose (often 75 g) is the easiest way to determine the glucose homeostasis status of human individuals (Box 10.2). Healthy persons have a fasting blood glucose level in the order of 5 mM, already 1 h after the glucose bolus show a peak below 10 mM and return to less than 7.8 mM after 2 h (Fig. 10.1, No. 1). Individuals that start at normal glucose concentrations but after 2 h still have levels higher than 7.8 mM have impaired glucose tolerance (Fig. 10.1, No. 2). However, when the fasting glucose level exceeds 7 mM and after 2 h still is higher than 11.1 mM, the person is considered diabetic (Fig. 10.1, No. 3). The response measured in the OGTT reflects (i) the ability of β cells to secrete insulin and (ii) the responsiveness of the whole body to insulin. For example, someone with a fasting glucose in the range of 6.1–7.0 mM is categorized to have impaired fasting glucose and may have established insulin resistance (Sect. 9.4). These individuals have impaired glucose homeostasis and are at increased risk to develop T2D.

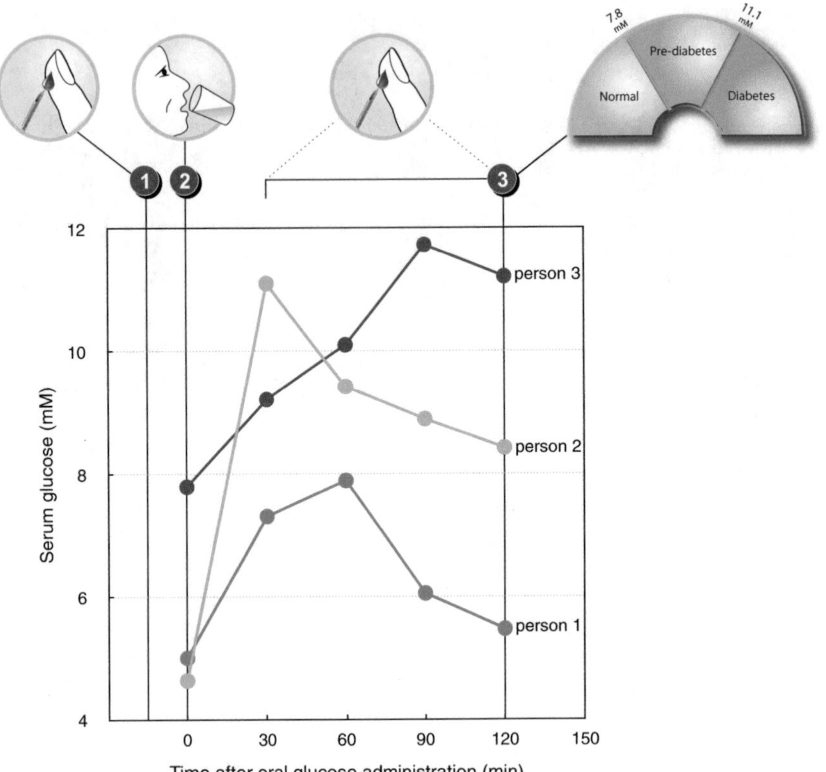

Fig. 10.1 Oral glucose tolerance test. The test measures how the human body responds to an oral challenge of glucose (usually as a drink of 75 g). Blood glucose is measured in a time course (for example, every 30 min over 2 h). The glucose level increases quickly, but the secretion of insulin should manage the normalization of the glucose concentration after 2 h (5 mM, person No. 1). Person No. 2 has normal fasting plasma glucose levels, but due to impaired glucose tolerance does not return after 2 h to normal concentrations (below 7.8 mM). In contrast, person No. 3 is diabetic, since his/her fasting glucose level already exceeds 7.8 mM, and the 2 h value is clearly elevated, respectively

The worldwide T2D prevalence of adults is 8.3 % (2014) and this rate will further increase (Fig. 10.2). The incidence of diabetes rises when countries become more industrialized, people eat a more sugar- and fat-rich diet and are less physical active. In high-income countries, primarily people above the age of 50 years become T2D, while in middle-income countries the highest prevalence is in younger persons. As these populations age, the prevalence will rise further due to the increase of older age groups. The mortality rate of diabetes varies sharply with the economy of the country. In 2011, the disease caused more than 3.5 million deaths in middle-income countries, of which more than 1 million were in China and just less than a million were in India. Mortality rates are significantly lower in high-income countries with a more developed healthcare system, but in 2011 also in the United States still some 180.000 people died due to T2D.

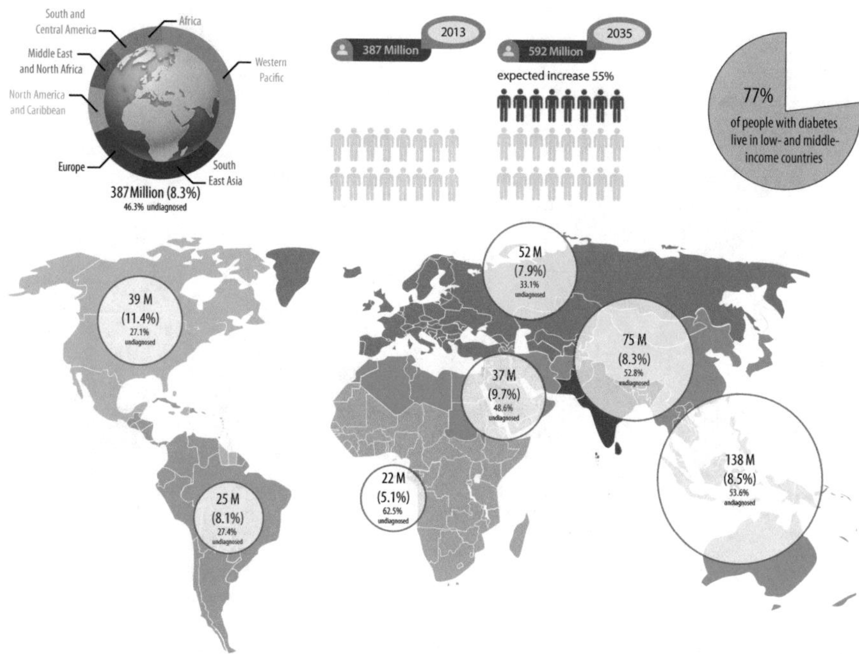

Fig. 10.2 Diabetes in numbers. The majority of the 387 million people with diabetes (2013) are between 40 and 59 years old. The present worldwide prevalence of the disease is 8.3 %. Until 2035, the number of people with diabetes will increase by 55 %. An additional 21 million cases of hyperglycemia during pregnancy, i.e. 17 % of live births to women in 2013, contribute to the global burden of diabetes. Worldwide diabetes caused 5.1 million deaths and consumed some 548 billion dollars (11 % of the total) in health spending in 2013. Despite the predominantly urban impact of the epidemic, T2D is rapidly becoming also a major health concern in rural communities in low- and middle-income countries. No country seems to be escaping this diabetes "epidemic"

10.2 Failure of Glucose Homeostasis in T2D and Its Treatment

T2D is an age-related disease that is strongly promoted by overnutrition and physical inactivity. In pre-disposed individuals, insulin secretion defects are mostly detected together with a reduced response to insulin-stimulated glucose uptake in the liver and adipose tissues, referred to as insulin resistance (Sect. 9.4). While the insulin resistance of a human individual remains relatively constant over time, the deterioration of the insulin-secretory capacity of β cells increases continuously. In the early stages of T2D, rising insulin levels maintain glucose tolerance by compensating for increased insulin resistance of skeletal muscles and adipose tissue. However, under these conditions, insulin is less potent in suppressing hepatic glucose production, i.e. also the liver becomes insulin resistant. In later stages of the disease, β cells lose their ability to compensate via the increase of insulin release resulting in reduced circulating insulin concentrations which often occurs in parallel

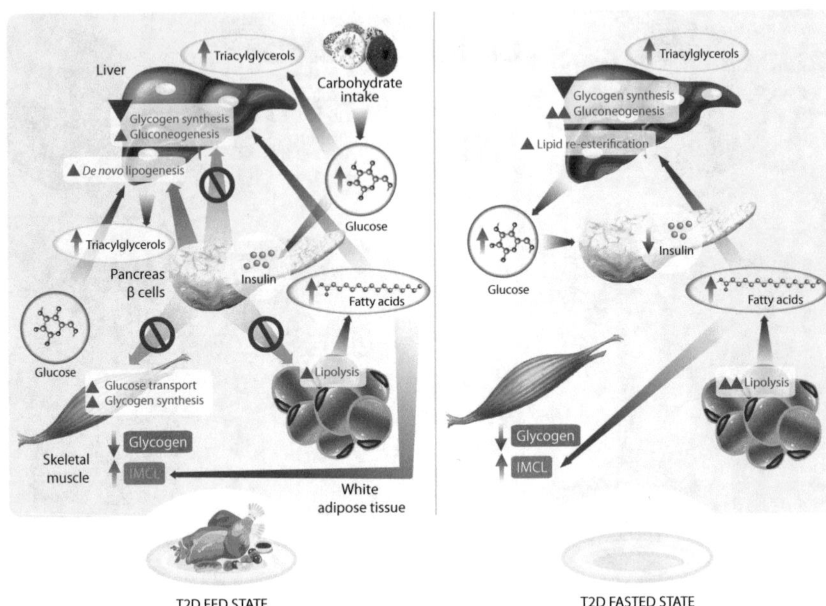

Fig. 10.3 Insulin action in T2D. In T2D, insulin-mediated skeletal muscle glucose uptake is impaired, which directs glucose to the liver. Increased hepatic lipid levels impair the ability of insulin to regulate gluconeogenesis and to stimulate glycogen synthesis. However, lipogenesis is not affected. In combination with increased delivery of dietary glucose, this stimulates lipogenesis causes NAFLD. Impaired insulin action in adipose tissue increases lipolysis, which promotes re-esterification of lipids in other tissues, such as the liver, and further exacerbates insulin resistance. In combination with a decline in the number of active β cells, this leads to the development of hyperglycemia. IMCL = intramyocellular lipid

with increased glucagon levels (Sect. 9.5). The shift in the glucagon/insulin ratio leads to a further rise in hepatic gluconeogenesis, i.e. more glucose is released by the liver to the circulation. In consequence, basal and post-prandial blood glucose levels are chronically increased (Fig. 10.1), i.e. the individual has hyperglycemia. Moreover, defective insulin secretion also causes dyslipidemia, including perturbed homeostasis of fatty acids, triacylglycerols and lipoproteins (Fig. 10.3) (Sect. 11.4).

The defective insulin secretion and responses in T2D have several reasons. For example, constant exposure of β cells to elevated levels of glucose and lipids, i.e. gluco-lipotoxicity, induces their dysfunction and ultimately triggers their death (Sect. 9.5). These processes are related to chronic inflammation of pancreatic islets (Fig. 10.4). Elevated glucose levels increase the metabolic activity of the islet cells, in which via increased ROS production, the NLRP3 inflammasome is activated (Sect. 7.1). In addition, increased insulin demand and production induces ER stress to β cells (Sect. 7.5) further activating the inflammasome. Moreover, endotoxins or Fetuin A-bound FFAs can stimulate TLR2 and TLR4 that lead to the translocation of NF-κB into the nucleus and the induction of the expression of pro-inflammatory genes, such as IL1B, IL6, IL8, TNF and CCL2. Like in other inflammatory scenarios (Sect. 7.4), this cytokine production leads to the attraction of macrophages and

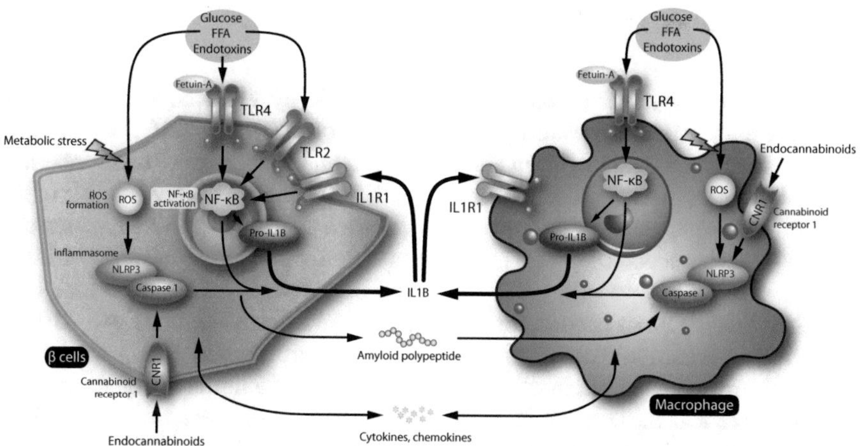

Fig. 10.4 β cells inflammation in T2D. Prolonged exposure of β cells to elevated concentrations of glucose and FFAs increases the metabolic activity of these cells and ROS formation. This activates the NLRP3 inflammasome, leading to the production and release of IL1B. Endocannabinoids can also activate the inflammasome via the cannabinoid receptor 1 (CNR1). In parallel, TLR2 and TLR4 are stimulated by endotoxins or Fetuin A-bound FFA, which via NF-κB activation causes the expression of pro-inflammatory genes. IL1B induces the expression of various cytokines and chemokines that lead to the attraction of macrophages. These macrophages are then activated by high levels of glucose, FFAs, endotoxins and endocannabinoids

other immune cells. Furthermore, the islets produce an amyloid polypeptide that aggregates to form amyloid fibrils in patients with T2D. In net result, resident islet macrophages adopt a pro-inflammatory M1 phenotype that induces islet dysfunction.

Current treatments for T2D include insulin, the indirect AMPK activator metformin, K^{ATP} channel inhibiting sulphonylureas, PPARγ-activating thiazolidinediones and inhibitors of either starch- and disaccharides-digesting α-glucosidase or of glucose transporters. Each of these therapies can improve hyperglycemia, and some may even delay the onset of diabetes. However, none of these drugs can slow down the progressive decline in insulin secretion. Intensive diabetes treatment results in tight glycemic control and therefore a substantial reduction in the risk of microvascular complications. Since T2D is often associated with hypertension and dyslipidemia (Sects. 11.1 and 11.4), respective drugs are prescribed to most patients with T2D in addition to glucose-lowering medications. However, several anti-diabetic drugs are associated with adverse effects, for example gastrointestinal symptoms with metformin therapy, increase in body weight with sulphonylureas, increase in body weight under insulin, and bone fractures with thiazolidinedione medication.

Importantly, T2D can be prevented by lifestyle changes. For example, already a moderately increase in physical activity combined with a decrease in caloric intake, aiming for a persistent 5–10 % weight loss, reduces the risk for T2D by more than 50 %. One goal of future personalized medicine should be to identify those patients who will most likely benefit from serious lifestyle changes, such as substantial weight loss.

10.3 Genetics of T2D

A range of monogenetic disorders result in chronic hyperglycemia (Table 10.1). They are summarized as maturity onset diabetes of the young (MODY), because they often occur already in young adults. However, the therapy of these inherited form of diabetes does not require insulin, i.e. they are of non-T1D type. Most of the MODY genes encode for transcription factors, such as *HNF4A*, *HNF1A*, pancreatic and duodenal homeobox 1 (*PDX1*), *HNF1B*, neuronal differentiation 1 (*NEUROD1*), *KLF11* and *PAX4*. In contrast *GCK*, carboxyl ester lipase (*CEL*) and B lymphoid tyrosine kinase (*BLK*) encode for enzymes and *INS* for a hormone. All MODY genes are expressed in β cells and affect insulin secretion, while the normal control of glucose metabolism via insulin involves a number of additional organs, such as muscle, liver and fat (Sect. 9.1). This suggests that insulin secretion in β cells is more important for diabetes than insulin resistance in peripheral organs.

However, in total the monogenetic forms of the disease represent only 1–2 % of diabetes cases worldwide. The remaining majority with typical obesity-related T2D is often found to carry a cluster of genetic variations that confer enhanced susceptibility to environmental factors, such as overnutrition, obesity and stress.

T2D belongs to those diseases that have been extensively studied by GWAS. Fig. 10.5 provides an overview on the 18 genetic variants that were the first

Table 10.1 MODY genes

Type	OMIM	Gene/protein	Description
MODY 1	125850	*HNF4A*	Due to a loss-of-function mutation in the *HNF4A* gene. 5–10 % of cases.
MODY 2	125851	*GCK*	Due to any of several mutations in the *GCK* gene. 30–70 % of cases. Mild fasting hyperglycemia throughout life. Small rise on glucose loading.
MODY 3	600496	*HNF1A*	Mutations of the *HNF1A* gene. 30–70 % of cases. Tend to be responsive to sulfonylureas. Low renal threshold for glucose.
MODY 4	606392	*PDX1*	Mutations of the *PDX1* gene. Less than 1 % of cases. Associated with pancreatic agenesis in homozygotes and occasionally in heterozygotes.
MODY 5	137920	*HNF1B*	Defect in *HNF1B* gene. 5–10 % of cases. Atrophy of the pancreas and several forms of renal disease
MODY 6	606394	*NEUROD1*	Mutations of the *NEUROD1* gene. Very rare.
MODY 7	610508	*KLF11*	Mutations of the *KLF11* gene.
MODY 8	609812	*CEL*	Mutations of the *CEL* gene. Very rare. Associated with exocrine pancreatic dysfunction.
MODY 9	612225	*PAX4*	Mutations of the *PAX4* gene.
MODY 10	613370	*INS*	Mutations in the *INS* gene. Usually associated with neonatal diabetes. Less than 1 % of cases.
MODY 11	613375	*BLK*	Mutated *BLK* gene. Very rare.

80 % of cases of early-onset, autosomal-dominant, familial hyperglycemia are represented by the 11 MODY genes

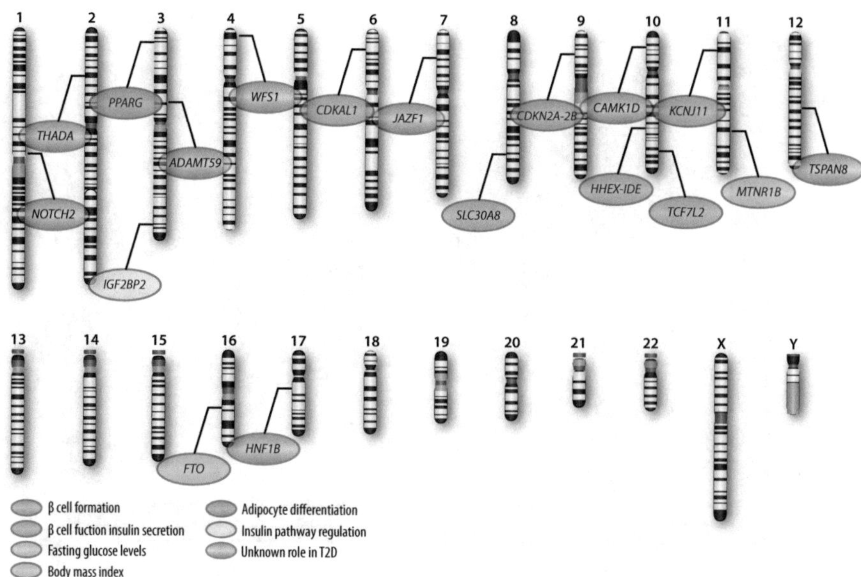

Fig. 10.5 Insights into the genetic basis of T2D. Examples of 18 genes are shown that were identified by GWAS to be associated with T2D. Four genes were previously known as T2D candidate genes, while the remaining 14 genes have previously not been suspected to play a role in T2D, such as *MTNR1B*, which shows the involvement of the circadian rhythms in T2D. The genes that participate in β cell disturbance have diverse functions, such as pancreatic islet proliferation, insulin secretion and cell signaling. SNPs of further six genes are statistically associated with T2D, but their role has not yet been identified

to be associated with T2D. They all represent common SNPs with minor allele frequencies (MAFs) ranging from 7.3 to 50 %. The gene transcription factor 7-like 2 (*TCF7L2*) has an odds ratio (OR) of 1.37 for developing T2D (i.e. an 37 % increased T2D risk), while the ORs for the 17 remaining genes range only between 1.05 and 1.15 (5–15 % increased risk). These numbers are comparable to what is observed for other common traits and diseases, such as obesity (Sect. 8.4). Some of the 18 T2D risk genes, such as CDK5 regulatory subunit associated protein 1-like 1 (*CDKAL1*), the zinc transporter *SLC30A8*, the transcription factor hematopoietically expressed homeobox (*HHEX*) and potassium inwardly-rectifying channel, subfamily J, member 11 (*KCNJ11*), are involved in insulin secretion in β cells. This suggests that the common risk for T2D agrees with the findings of monogenetic diabetes.

However, the genetic propensity to develop T2D involves also genes in a number of additional pathways that affect β cell formation and function, as well as pathways affecting fasting glucose levels and obesity. The cluster of the cyclin-dependent kinase inhibitor (CDKN) 2A and 2B controls β cell growth, melatonin receptor 1B (*MTNR1B*) links circadian rhythms (Sect. 3.6) with fasting glucose levels, *FTO* is the key risk gene for obesity (Sect. 8.4), *PPARG* encodes for the master regulator of adipogenesis (Sect. 8.5) and insulin-like growth factor 2 mRNA binding protein 2

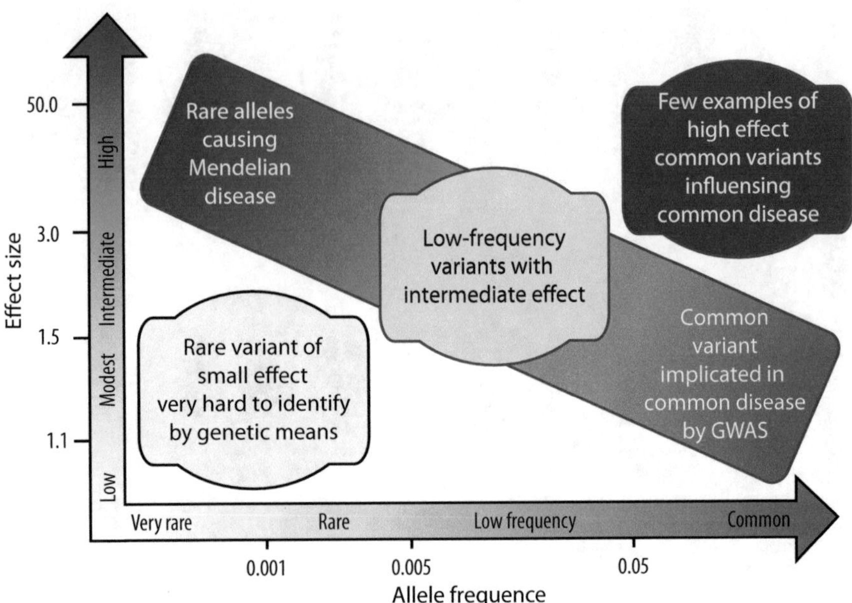

Fig. 10.6 Identifying genetic variants by risk allele frequency. The strength of the genetic effect is indicated by odds ratios. Most emphasis and interest lies in identifying associations with characteristics shown within the diagonal box

(*IGF2BP2*) is involved in insulin signaling (Sect. 9.2). However, for the remaining T2D risk genes no direct link to metabolic homeostasis had been identified so far (Fig. 10.5). This indicates that not always the closest genes to the T2D-associated SNP provide a functional explanation, but in some cases more distant genes may be involved.

However, the 18 genetic variants explain only some 4 % of the heritable risk for T2D. In the meantime, larger study populations and meta-analyses of existing studies increased the number of T2D risk genes to more than 40 (www.genome.gov/gwastudies). The additional genes have similar or even lower MAFs and ORs. Nevertheless, in personalized medicine approaches, such as exemplified iPOP (Sect. 4.6), the presently known T2D risk genes are already used as predictive markers. Moreover, for some forms of MODY different therapies are suggested, such as high sensitivity of *HNF1A* mutations to sulphonylurea drugs. Taken together, this suggests that T2D is a very heterogenous disease that can be segregated into multiple subtypes, which should be treated on a personalized basis on the individual's genetic background and phenotype.

In general, common SNPs are characterized by low ORs, while rare monogenetic forms of T2D have high ORs (Fig. 10.6). However, both extremes do not explain all genetic basis of T2D. Like for many other common diseases and traits, also for T2D it is assumed that there is a large number of low frequency SNPs with intermediate ORs. These genetic variants will be identified by the use of whole genome sequencing (Sect. 2.6). Nevertheless, these additional genetic variations will not be able to explain all heritability of T2D. In contrast, pre-natal and post-natal epigenetic programming via DNA methylation (Sect. 10.4) will demonstrate its contribution to the disease risk.

10.4 Thrifty Gene Hypothesis

Some human populations, such Polynesians, show an overproportional high prevalence for T2D. One likely explanation is that their ancestors experienced a number of challenges, such as cold stress and starvation, during long open-ocean travels over the Pacific, which often let only the initially most obese members of the group survive. Therefore, these ancestors may have been evolutionary selected for energetic efficiency, referred to as "thrifty" metabolism. The thrifty gene hypothesis (Fig. 10.7), proposed already 50 years ago, suggests that present-day Polynesians have inherited an increased T2D susceptibility because their ancestors went through an evolutionary "bottleneck". A similar explanation may apply to a number of other indigenous populations, such as the Pima Indian tribe in Arizona who had adapted to deprivation of life in the desert. For example, non-diabetic Polynesians and Pima Indians have post-prandial insulin plasma levels that are 3-times higher than those of Europeans. Similarly, members of both populations get far more likely obese than Europeans, when they eat the same amount of calories. Since obesity is a dominant risk factor for diabetes, these individuals then get far more likely T2D. The

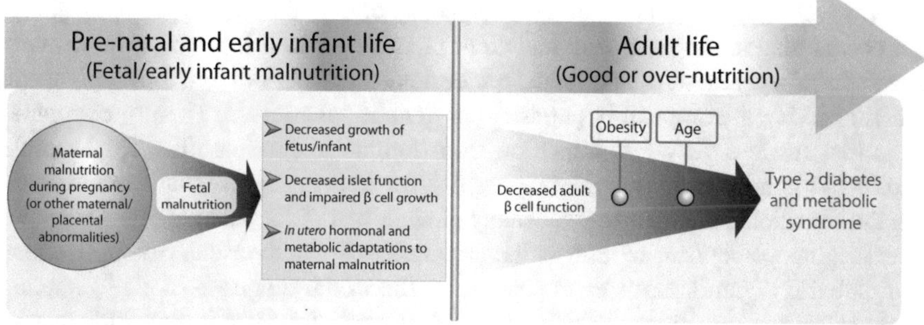

Fig. 10.7 Thrifty gene hypothesis in T2D. T2D can be the outcome of maternal-fetal malnutrition, such as poor maternal diet, low maternal fat stores or reduced transfer of nutrients because of placental abnormalities. The fetus adapts to this environment by being nutritionally thrifty, resulting in decreased fetal growth, islet function and β cell mass and other hormonal and metabolic adaptations. A transition to overnutrition later in adult life exposes the impaired islet function to increased metabolic stress, which is further enhanced by obesity and age, so that T2D results. Thus, the thrifty gene hypothesis is based on an altered epigenetic programming that aids survival pre- and early post-natally

more a human population have been selected for longer periods of famine in recent evolutionary history, the more their members are prone to develop obesity and T2D when exposed to high-fat diet. In the past it was a clear evolutionary advantage, when an individual could store efficiently energy in form of WAT hypertrophy, since food was never constantly available in large amounts. In contrast, in today's obesogenic environment, the offspring of these people suffers from obesity and T2D. This also implies that those human populations who did not experience long periods of starvation in past few 100 years, such as the Europeans, have still a relatively low prevalence for T2D. Interestingly, the prevalence of T2D in European immigrants of British and German ancestry to the United States and Australia is significantly higher than that of British and German people still living in Europe with similar lifestyles. The people who stayed in Europe tended to be richer than those who emigrated, i.e. poorer people are more likely starvation-prone carrying the thrifty genotype.

The experience from the already discussed Dutch hunger winter (Sect. 5.5) provides a molecular explanation for an increased obesity and T2D risk. The *in utero* environment has a strong impact on fetal programming, as individuals exposed to either famine or maternal gestational diabetes during fetal development develop obesity and/or T2D later in life more likely. Food-deprived conditions of the mother change the epigenome of the fetus, so that genes involved in energy homeostasis are more sensitive to food intake. In times of continued famine, this epigenetic sensitizing is a survival advantage, while in times of plenty food it may drive the individual into obesity and T2D. DNA methylation is particularly sensitive to events during development *in utero*, because DNA is almost fully demethylated during zygote formation, and specific methylation is re-established throughout embryogenesis. Patterns of DNA methylation at candidate genes, such as the *LEP* gene, associate with *in utero* exposure to famine and to maternal impaired glucose tolerance during pregnancy.

However, epigenetic programming happens not only during the pre-natal phase, but occurs also in the post-natal and adult phases of life. It is not clear how many generations' epigenetic marks can be inherited, but it is obvious that epigenetic inheritance is not comparably persistent as genetic inheritance. Therefore, populations that made a very fast transition from famine to food surplus, for example within 1–2 generations, are under higher risk for T2D than those that were improving the nutritional conditions over many generations. T2D patients, who maintain intense glucose control, remain at increased risk of macrovascular complications and diabetic organ damage even years after the initial diagnosis. This "glycemic memory" is also an epigenetic effect, i.e histone methylation changes within human aortic endothelial cells in response to increased glucose exposure.

Future View

Diabetes, like cancer, is a very heterogenous disease that in future will be diagnosed and treated on the basis of their molecular features, i.e. on a far more personalized basis. One particular goal will be to use personalized lifestyle and medication, in order to reduce the risk of cardiovascular complications of T2D. Since most cases of T2D are preventable, in future far more effort will be taken in detecting early genetic and epigenetic markers for increased T2D susceptibility. Such epigenetic markers will help to detect children, or families, respectively, for whom intensive lifestyle intervention are likely to prevent the onset of metabolic disease.

Key Concepts

- Diabetes is a condition of chronically elevated plasma glucose levels that eventually causes toxicity to blood vessels.
- In contrast to people with T1D, T2D patients at least in the beginning do not necessarily require daily doses of insulin to survive.
- Healthy persons have a fasting blood glucose level in the order of 5 mM, already 1 h after a defined glucose bolus show a peak with less then 10 mM and return to less than 7.8 mM after 2 h. However, when the fasting glucose level exceeds 7 mM and is after 2 h still higher than 11.1 mM, the person fulfills the definition of T2D.
- The worldwide T2D prevalence of adults is 8.3 % (2014) and will further rise.
- T2D basically is an age-related disease that is strongly promoted by overnutrition and physical inactivity. In pre-disposed individuals, insulin secretion defects are mostly detected together with a reduced response to insulin-stimulated glucose uptake in the liver and adipose tissues, referred to as insulin resistance.
- All MODY genes are expressed in β cells and affect insulin secretion, while the normal control of glucose metabolism via insulin involves a number of additional organs, such as muscle, liver and fat.
- T2D is a very heterogenous disease that can be divided into multiple subtypes, which might be treated personalized based on the individual's genetic background and phenotype.
- The thrifty gene hypothesis suggests that the ancestors of some human populations have an epigenetically programmed increased T2D susceptibility because their ancestors went through an evolutionary bottleneck of prolonged periods of starvation. The more a human population had been selected for longer periods of famine in recent evolutionary history, the more their members are prone to develop obesity and T2D when exposed to the high-fat diet.
- Populations that made a very fast transition from famine to food surplus, such as within 1–2 generations, are under higher risk for T2D than those that had improved their nutritional conditions over many generations.

Additional Reading

Donath MY (2014) Targeting inflammation in the treatment of type 2 diabetes: time to start. Nat Rev Drug Discov 13:465–476

Grayson BE, Seeley RJ, Sandoval DA (2013) Wired on sugar: the role of the CNS in the regulation of glucose homeostasis. Nat Rev Neurosci 14:24–37

International Diabetes Federation (IDF): www.idf.org/diabetesatlas

Laakso M, Kuusisto J (2014) Insulin resistance and hyperglycaemia in cardiovascular disease development. Nat Rev Endocrinol 10:293–302

Manolio TA, Collins FS, Cox NJ, Goldstein DB, Hindorff LA, Hunter DJ, McCarthy MI, Ramos EM, Cardon LR, Chakravarti A et al (2009) Finding the missing heritability of complex diseases. Nature 461:747–753

Reddy MA, Zhang E, Natarajan R (2015) Epigenetic mechanisms in diabetic complications and metabolic memory. Diabetologia 58:443–455

Chapter 11
Hypertension, Atherosclerosis and Dyslipidemias

Abstract Hypertension is the most important preventable risk factor for pre-mature death. Chronically elevated blood pressure increases the risk of ischemic heart disease, stroke, peripheral vascular disease and other CVDs, such as heart failure, aortic aneurysms, diffuse atherosclerosis and pulmonary embolism. The high consumption of saturated fat, low intake of fruit and vegetables and whole grain fiber high-cholesterol diets, i.e. Western-type diet, can lead to hypercholesterolemia and atherosclerosis, especially in genetically predisposed individuals. Atherosclerosis is a chronic inflammatory disease caused by the accumulation of cholesterol-laden macrophages in the artery wall, i.e it is based on dyslipidemia and an overeaction of the immune system. Central cells in atherosclerosis are macrophages and their phenotype, i.e. their programming to M1 and M2 type, which influences both disease progression and regression.

In this chapter, we will link three important risk factors for heart disease, hypertension, atherosclerosis and dyslipidemia, in a combined mechanistic model. We will understand that the susceptibility to CVD is associated with levels of plasma lipids and lipoproteins. Like in obesity and T2D, the genetic basis of the cardiometabolic traits, such as plasma lipoprotein levels, can be explainable to a minor part on a monogenetic basis causing a large effect and to a larger part by common genetic variants with minor effects on the trait. This will provide another example how molecular medicine is applied, in order to diagnose and treat the basis of a disease.

Keywords Hypertension • CHD • Stroke • Renal disease • Atherosclerosis • Chronic inflammation • Foam cells • ER stress • LXR • Cholesterol • Triacylglycerols • Lipoproteins • LDL • HDL • Apolipoproteins • Dyslipidemia

11.1 Hypertension

Each cycle of heart contraction pumps some 70 ml blood into the systemic arterial system, in order to supply all cells and tissues of the human body with oxygen and nutrients. This pulsation creates pressure on the vascular walls that depends on the amount of pumped blood and the resistance created by the vasculature. Due to the periodic ejection of blood from the heart, this pressure is highest at the peak of a

C. Carlberg et al., *Nutrigenomics*, DOI 10.1007/978-3-319-30415-1_11

Box 11.1 Healthy and Unhealthy Blood Pressure Ranges

Blood pressure varies during each heartbeat between a maximum (referred to as systolic) and a minimum (diastolic) value (Table 11.1). The pressure displays a circadian rhythm with highest values in the afternoon and lowest at night. The mean blood flow depends on the blood pressure and the resistance to flow caused by the mechanics of the blood vessel walls, i.e. it decreases with distance from the heart, in particular due to narrowing of small arteries and arterioles. Blood pressure is also affected by hydrostatic forces, such as during standing (orthostasis), valves in veins, breathing and skeletal muscle contraction, respectively.

Table 11.1 Blood pressure ranges. The systolic blood pressure indicates how much pressure blood is exerting against the artery walls while the heart is pumping blood, while the diastolic blood pressure measures the pressure while the heart is resting between beats. The ranges for normal blood pressure, pre-hypertension, hypertension (stages I and II) and hypertensive crisis are defined. In individuals older than 50 years, hypertension is present when the blood pressure is consistently above 140 mm Hg systolic or 90 mm Hg diastolic.

Blood pressure category	Systolic mm Hg (upper level)		Diastolic mm Hg (lover level)
Normal	less then 120	and	less than 80
Pre-hypertension	120–139	or	80–89
Hypertension, stage 1	140–159	or	90–99
Hypertension, stage 2	160 or higher	or	100 or higher
Hypertensive crisis	higher than 180	or	higher than 110

passing amount of blood (systolic) and lowest after its passage (diastolic). For healthy adults the values should be in the order 120 millimeters of mercury (mm Hg) systolic and 80 mm Hg diastolic, respectively (Box 11.1 and Table 11.1). Blood pressure varies with physical activity and depends on diseases, such as obesity. It is tightly regulated by signals from the SNS, in order to permit uninterrupted blood perfusion of all vital organs. For example, even transient interruption in blood flow to the brain causes loss of consciousness, and longer interruptions will result in death of unperfused tissues, such as in cerebral stroke or myocardial infarction (Sect. 11.2).

Increased salt reabsorption in the kidneys, for example after a salty meal, requires increased water reabsorption, in order to maintain plasma sodium concentration at some 140 mM. This results in an increased intravascular volume boosting venous blood return to the heart. Thus, the cardiac output raises and leads to elevated blood pressure. A central point for salt absorption is the Na^+ channel SCNN1 in the cortical collecting tubule of the kidney. Decreased delivery of salt leads to increased secretion of the aspartyl protease renin resulting in the secretion of the peptide hormone angiotensin. This causes vasoconstriction and a subsequent increase in blood pressure. Angiotensin binds to its specific G protein-coupled receptor in the adrenal gland inducing there the production of aldosterone, i.e. the ligand of the nuclear receptor mineralocorticoid receptor (MR). MR then up-regulates *SCNN1* gene expression leading to an increase of salt re-absorption. In addition to this renin-angiotensin-aldosterone system, there are (i) baroreceptors sensing acute changes in the pressure of blood vessels, (ii) natriuretic peptides produced by the brain and heart in response to increased blood pressure in these organs, (iii) the kinin-kallikrein system affecting vascular tone and renal salt handling, (iv) the adrenergic receptor system influencing heart rate, cardiac contraction and vascular tone and (v) factors produced by blood vessels causing vasodilation, such as NO, or contraction, such as endothelin. These control systems work in an integrated fashion, in order to ensure adequate perfusion of all tissues despite widely varying metabolic demand.

Hypertension is defined as the blood pressure level above which therapeutic intervention has clinical benefit. The most common form of hypertension, accounting for 90–95 % of all cases, is primary (also referred to as "essential") hypertension that results from a complex interaction of genes and environmental factors. Chronic hypertension in combination with atherosclerosis (Sect. 11.2) is the major risk factor for stroke, CHD, congestive heart failure and end-stage renal disease (Fig. 11.1). Obesity can increase the risk of hypertension fivefold as compared with normal weight, and more than 85 % of hypertension cases can be attributed to a BMI > 25.

Like obesity and T2D, hypertension has been intensively investigated by GWAS and more than 30 common SNPs with small effects on blood pressure as well as some rare genetic variants with large effects are known. The latter converge on a common pathway of altering blood pressure by changing net renal salt balance. This emphasizes salt homeostasis in the kidney as a key risk factor for hypertension. In fact, approximately one third of hypertensive persons are responsive to sodium intake. From an evolutionary perspective this makes sense, since humans originated from a notoriously salt-poor environment of sub-Saharan Africa providing a strong adaptive advantage for gene variants that promoted salt and water retention. After migrating to salt-rich environments, such as Europe, the same variants now contribute to elevated blood pressure and its health consequences. Thus, dietary factors, such a salt intake, significantly influence blood pressure and reduced dietary salt intake as well as increased consumption of fruits and low fat food, exercise, weight loss and reduced alcohol intake can reduce hypertension.

11.2 Mechanisms of Atherosclerosis

The endothelium is a single layer of endothelial cells that covers blood vessels and serves as a barrier between the circulating blood and subendothelial tissues. In atherosclerosis, cholesterol deposition below the endothelium causes a macrophage-dominated inflammatory response in large and medium arteries. Some atherosclerotic lesions can develop already in the first years of life, and 95 % of humans by the age of 40 have some type of lesion. However, in most cases clinical manifestations

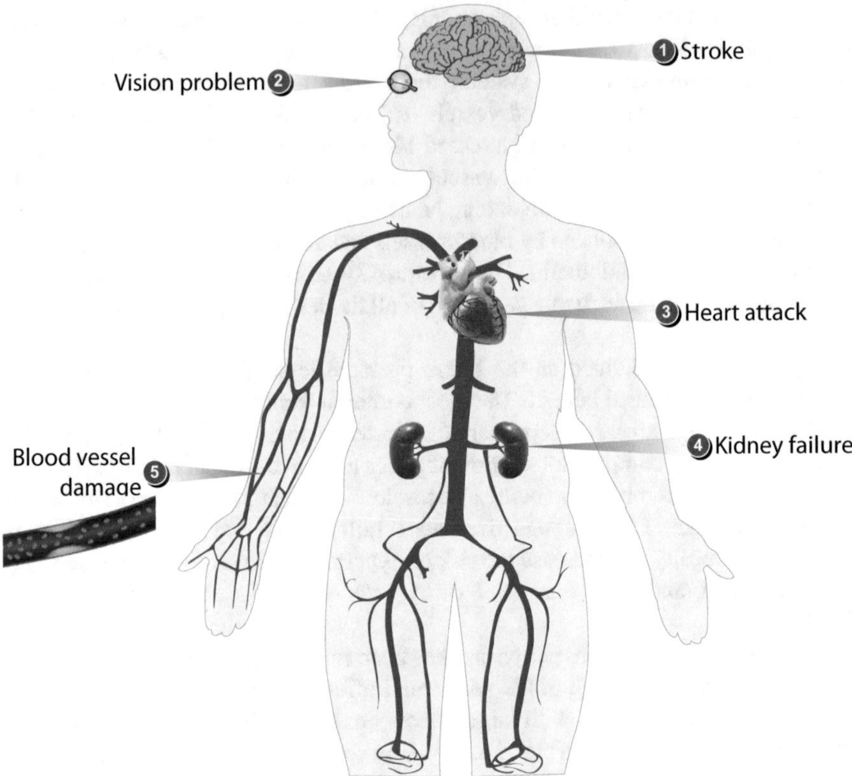

Fig. 11.1 Main complications from hypertension. Hypertension is the most important preventable risk factor for pre-mature death worldwide. It increases the risk of ischemic heart disease, strokes, peripheral vascular disease and other cardiovascular diseases, including heart failure, aortic aneurysms, diffuse atherosclerosis and pulmonary embolism. Hypertension is also a risk factor for cognitive impairment and dementia as well as for chronic kidney disease. Other complications include hypertensive retinopathy and hypertensive nephropathy

occur not before the age of 50–60 years. The distribution of atherosclerotic plaques, i.e. the core lesions of atherosclerosis, is not randomly, but they tend to accumulate at the inner curvatures and branch points of arteries, i.e. at positions where laminar flow is either disturbed or insufficient, in order to maintain the normal, quiescent state of the endothelium.

A first sign of lesion formation is the accumulation of cholesterol within the arterial wall, referred to as fatty streaks. This is initiated by the sequestration of cholesterol-rich, APOB-containing lipoproteins, called LDLs (Sect. 11.3). When LDLs are modified by oxidation, enzymatic and non-enzymatic cleavage and/or aggregation, they become pro-inflammatory and stimulate endothelial cells to produce chemokines, such as CCL5 and CXCL1, glycosaminoglycans and P-selectin for the recruitment of monocytes. Hypercholesterolemia and cholesterol accumulation in hematopoietic stem cells promotes the overproduction of monocytes leading

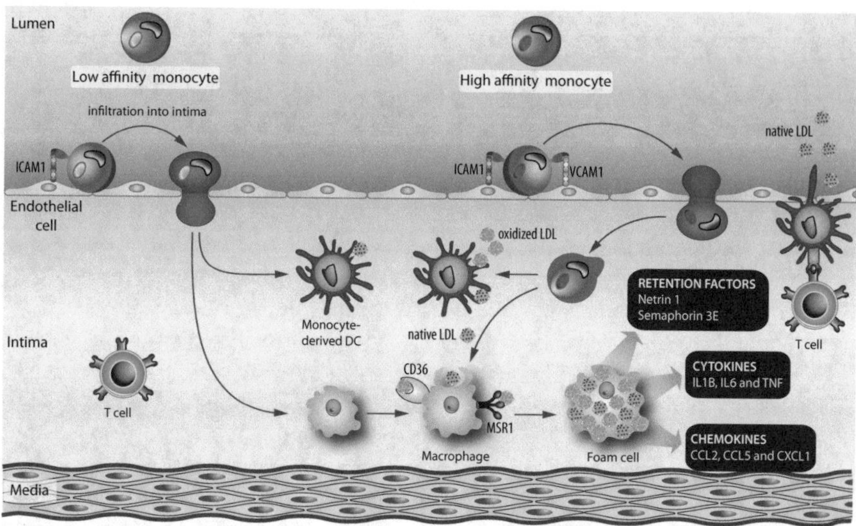

Fig. 11.2 Monocyte recruitment and accumulation in plaques. Hyperlipidemia increases the number of monocyte subsets that are recruited to atherosclerotic plaques. Different chemokine-chemokine receptor pairs and endothelial adhesion molecules, such as selectins, ICAM1 and VCAM1, mediate the infiltration of the monocytes into the intima. There the monocytes differentiate into macrophages or dendritic cells and interact with atherogenic lipoproteins. Macrophages take up native and oxidized LDLs via macropinocytosis or scavenger receptors, such as MSR1 and CD36, resulting in the formation of foam cells (Box 11.2). These foam cells secrete pro-inflammatory cytokines, such as IL1B, IL6 and TNF, and chemokines, such as CCL2, CCL5 and CXCL1, as well as macrophage retention factors, such as netrin 1 and semaphorin 3E, that amplify the inflammatory response

to increased adherences to endothelial cells. This is mediated by the molecules integrin, alpha 4 (ITGA4) and integrin, beta 2 (ITGB2) in monocytes binding to vascular cell adhesion molecule 1 (VCAM1) and intercellular adhesion molecule 1 (ICAM1) in endothelial cells (Fig. 11.2). The monocytes then move into the space below the endothelial cells, referred to as "intima", where they differentiate into macrophages. These macrophages ingest normal and modified lipoproteins, such as oxidized LDL, which at the onset of the inflammatory response is a beneficial clearance. Hypercholesterolemia (Sect. 11.4) is important for the recruitment of macrophages into the arterial wall, but also immunological and mechanical injuries, as well as bacterial and viral infections, contribute to the pathogenesis of atherosclerosis. When the inflammation turns chronic, the macrophages transform into cholesterol-laden foam cells (Box 11.1). These cells persist in plaques, i.e. they have lost their ability to migrate and cannot resolve inflammation.

Many different cell types, such as endothelial cells, monocytes, dendritic cells, lymphocytes, eosinophils, mast cells and smooth muscle cells, contribute to the process of atherosclerotic plaque formation, but foam cells are central to the pathophysiology of the disease. The TLR-dependent activation of these monocyte-derived cells polarizes them to M1 macrophages (Sect. 7.1) that secrete pro-atherosclerotic

Box 11.2 Foam Cells

Atherosclerosis is both a lipid disorder and an inflammatory disease, in which macrophages play a central role. One of the earliest pathogenic events in the nascent plaque is lipoprotein uptake by monocyte-derived macrophages. This results in the development of foam cells (Fig. 11.3). Increased oxidative stress in the artery wall promotes modifications of LDLs, which are primarily oxidations that are recognized by macrophages via scavenger receptors, such as MSR1 and CD36. These receptors internalize the lipoproteins and cholesteryl esters of the lipoproteins are hydrolyzed to free cholesterol and fatty acids. Importantly, the scavenger receptors are not down-regulated by cholesterol via a negative-feedback mechanism, such as in the case of LDLR. The free cholesterol is transported to the ER, where it is re-esterified by the enzyme acetyl-CoA acetyltransferase 1 (ACAT1) providing the "foam" in foam cells. In addition, a number of other mechanisms, such pinocytosis by macrophages, contribute to foam cell formation. Enrichment of ER membranes with free cholesterol can result in defective cholesterol esterification by ACAT1 in macrophages, promoting the accumulation of free cholesterol. Moreover, free cholesterol enrichment in lipid raft within the cell membranes can enhance inflammatory signaling via TLR4 and NF-κB. In addition, atherosclerotic plaques contain cholesterol crystals that induce the NLRP3 inflammasome (Sect. 7.1) and promote apoptosis of the foam cells. Importantly, the transporter proteins ABCA1 and ABCG1 are able to mediate the reverse cholesterol transport (Sect. 7.3). ABCA1 promotes cholesterol efflux to lipid-poor APOA1 of HDLs, whereas ABCG1 stimulates the efflux to mature HDLs. Since the genes *ABCA1* and *ABCG1* are regulated by the oxysterol sensor LXR (Sect. 3.4), synthetic LXR agonists have a high potential for the treatment of atherosclerosis. Another therapeutic strategy would be to polarize macrophages to the M2 phenotype that would regress atherosclerotic plaques via the secretion of anti-inflammatory cytokines, such as IL10, promote tissue remodeling and clear of dying cells and debris.

cytokines, such as IL6 and IL12, matrix-degrading proteases as well as ROS and NO. The plaque macrophages are subject to both retention and emigration signals and a misbalance of these processes contributes to the net accumulation of plaque macrophages. Dysregulation of lipid metabolism in foam cells contributes to ER stress ultimately resulting in apoptotic cell death. Since defective lipid metabolism pathways, such as cholesterol esterification and efflux, impair efficient clearance of apoptotic cells, this results in necrosis and in the release of cellular components and lipids that form the necrotic core (Fig. 11.4). Chronic inflammation stimulates the migration of smooth muscle cells into the intima region, where they transform into fibroblasts that proliferate and produce larger amounts of extracellular matrix. This leads to the formation of fibrous atherosclerotic plaques. Due to the calcification of the plaques, the artery wall becomes rigid, i.e. sclerotic and fragile. Most humans

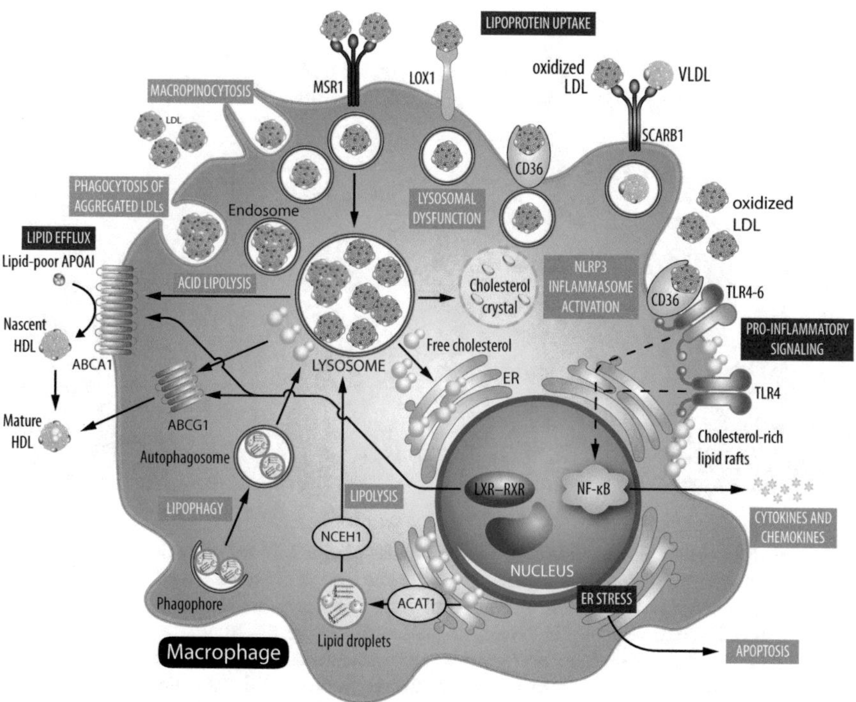

Fig. 11.3 Lipoprotein uptake and efflux in macrophages. Via macropinocytosis, phagocytosis of aggregated LDLs and scavenger receptor-mediated uptake, macrophages within the plaque internalize native LDLs and VLDLs as well as oxidized lipoproteins. The internalized lipoproteins and their associated lipids are digested in the lysosome releasing free cholesterol. The latter moves to the plasma or ER membrane and can be effluxed from there. The enzyme ACAT1 in the ER membrane esterifies the free cholesterol with fatty acids for their storage within cytosolic lipid droplets. Via lipophagy resulting in the delivery of lipid droplets to lysosomes for efflux of via lipolysis by neutral cholesterol ester hydrolase 1 (NCEH1), these lipids can be mobilized. LXR is activated by the accumulation of cellular cholesterol and up-regulates the expression of *ABCA1* and *ABCG1*. These genes mediate the transfer of free cholesterol to lipid-poor APOA1 to form nascent HDLs or mature HDLs, in which free cholesterol is esterified and stored in the core of the particle. Cholesterol crystal formation in the lysosome is stimulated by excessive free cholesterol accumulation and can activate the NLRP3 inflammasome, promotes ER stress and leads to apoptosis. Moreover, lipid rafts are enriched in sphingomyelin form a complex with free cholesterol and promote TLR4 signaling. This leads to activation of NF-kB and in the production of pro-inflammatory cytokines and chemokines

have small atherosclerotic lesions that do not compromize blood flow. However, in subclinical atherosclerosis some lesions reduce the oxygen supply of tissues. The growth of the lesions and inward arterial remodeling gradually narrow the diameter of the blood vessels and thus reduces the blood flow (referred to as stenosis). When this stenosis applies to more than 80 % of the coronary arteries, the heart muscle becomes ischemic, especially when a high cardiac workload increases oxygen demand. Finally, the originally stable lesion changes into an unstable vulnerable

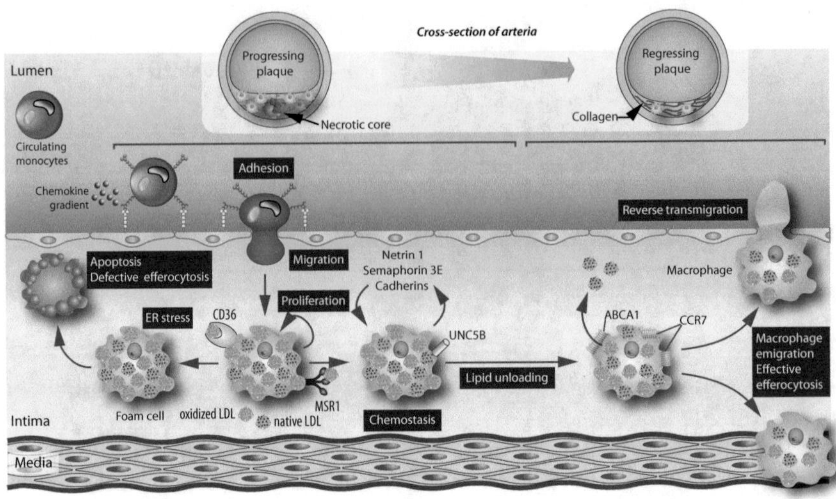

Fig. 11.4 Macrophage retention and emigration in plaques. Imbalances in the lipid metabolism of macrophages within the progressing plaque can lead to the retention of the cells and subsequently to chronic inflammation. Retention molecules, such as netrin 1 and its receptor unc-5 homolog B (UNC5B), semaphorin 3E and cadherins, are expressed by lipid-laden macrophages express and promote macrophage chemostasis. This inflammatory milieu causes ER stress, leads to apoptosis and results in the formation of the necrotic core. Lipid unloading via ABCA1 and cholesterol efflux can reverse foam cell accumulation (Fig. 11.3). In parallel, in myeloid-derived cells the chemokine receptor CCR7 is up-regulated and the expression of retention factors is decreased

plaque that can easily rupture the endothelium leading to the formation intravascular blood clots that can cause myocardial infarction or, in the case of damage of brain supplying arteries, in cerebral stroke.

11.3 Cholesterol Metabolism and Lipoproteins

Cholesterol is essential for membrane structure and fluidity and is a precursor to steroid hormones, vitamin D_3, oxysterols and bile acids that are ligands of nuclear receptors (Sect. 3.2). Only a small amount of circulating cholesterol originates from nutrition, while approximately 80 % derives from endogenous synthesis. Cholesterol levels are tightly regulated by the coordinated actions of the transcription factors SREBF1 (Sect. 3.1) and LXR (Sect. 3.4). At low cholesterol levels, SREBF1 activates genes that are involved in endogenous cholesterol production and cholesterol uptake, such as *LDLR*. Since already low concentrations of free cholesterol can be toxic, cholesterol is mostly esterified with fatty acids. Cholesterol and cholesteryl esters are insoluble in plasma, which requires their transport in spheroidal macromolecules, referred to as lipoproteins. These lipoproteins have a hydrophobic core formed by phospholipids, fat-soluble anti-oxidants, vitamins and cholesteryl esters and a hydrophilic coat that contains free cholesterol, phospholipids and apolipoproteins.

There are four main types of lipoproteins: chylomicrons, VLDLs, LDLs and HDLs, which are differentiated based on their density and size (Box 11.3). The density of lipoproteins depends on the abundance of apolipoproteins. Chylomicrons, VLDLs and in particular LDLs deposit cholesterol in peripheral tissues. Increased levels of cholesterol-rich LDLs are associated with elevated risk of CVD. LDLs carry preferentially APOB and can deliver cholesterol to artery walls leading to the formation of atherosclerotic plaques (Sect. 11.2). HDLs have a high amount of APOA1 and mediate the reverse cholesterol transport to the liver (Sect. 7.3). Thus, in contrast to LDLs, high levels of HDLs, sometimes referred to as "good cholesterol", are associated with reduced risk for CVD. The ratio of APOB to APOA1 is the strongest predictor of CHD risk, but the ratio of total cholesterol to HDL-cholesterol is equally predictive. For example, an increase of total cholesterol of 5.2–6.2 mM is associated with a threefold elevated risk of death from heart attack, while a HDL-cholesterol level lower than 0.9 mM is the most common lipid disturbance of patients below the age of 60. Fig. 11.5 provides an overview on the lipoprotein metabolism.

In contrast to most other chronic inflammatory diseases, in atherosclerosis there is the potential to remove the inflammatory stimulus. Lowering of plasma LDL levels by drugs, such as statins that target HMGCR (the rate-limiting enzyme in the production of cholesterol) can prevent subendothelial retention of lipoproteins and thereby decrease inflammatory atherosclerotic disease. The enzyme proprotein convertase subtilisin/kexin type 9 (PCSK9) binds to LDLR and induces its degradation, which reduces the metabolism of LDL-cholesterol and can lead to hypercholesterolemia. Therefore, inhibitory antibodies against PCSK9 were developed and are

Box 11.3 Lipoproteins

The composition of the four types of lipoproteins is listed.

Chylomicrons: With 50–200 nm diameter they represent the largest lipoproteins and show low density (<1.006 g/ml). They are composed of approx. 85 % triacylglycerols, 9 % phospholipids, 4 % cholesterol and 2 % protein, such as APOB48.

VLDLs: Very low density (0.95–1.006 g/ml) lipoprotein of 30–70 nm diameter containing approx. 50 % triacylglycerols, 20 % cholesterol, 20 % phospholipids and 10 % protein, such as APOB100.

LDLs: Low density (1.016–1.063 g/ml) lipoproteins of 20–25 nm diameter that are composed of approx. 45 % cholesterol, 20 % phospholipids, 10 % triacylglycerols and 25 % protein, such as APOB.

HDLs: High density (1.063–1.210 g/ml) lipoproteins with a diameter of 8–11 nm that are formed of approx. 40–55 % protein, such as APOA1, 25 % phospholipids, 15 % cholesterol and 5 % triacylglycerols.

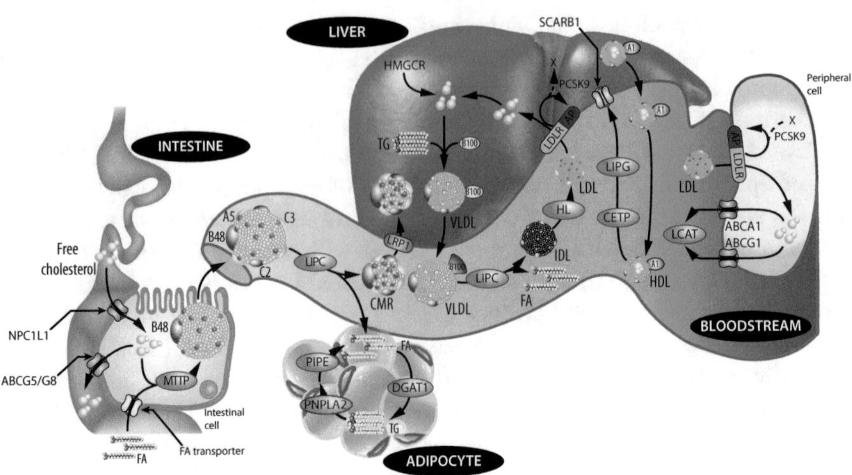

Fig. 11.5 Metabolism of triacylglycerols, LDL-cholesterol and HDL-cholesterol. The main lipids in lipoproteins are free and esterified cholesterol and triacylglycerols. In triacylglycerol metabolism, hydrolyzed dietary fats enter intestinal cells via fatty acid transporters. Through a vesicular pathway microsomal triglycerole transfer protein (MTTP) packs reconstituted triacylglycerols with cholesteryl esters and APOB48 into chylomicrons. Chylomicrons also contain the apoliproteins APOA5, APOC2 and APOC3. In adipocytes, the enzyme diacylglycerol O-acyltransferase 1 (DGAT1) re-synthesizes triacylglycerols that had been hydrolyzed by PNPLA2 and hormone sensitive lipase (LIPE). Chylomicron remnants are taken up by hepatic LDLR or LDLR-related protein 1 (LRP1). In liver cells, triacylglycerols are packaged with cholesterol and APOB100 into VLDLs. The triacylglycerols in VLDLs are hydrolyzed by LPL, which releases fatty acids and VLDL remnants (IDLs) that are hydrolyzed by LIPC yielding LDLs. Sterols in the intestinal lumen enter intestinal cells via the transporter Niemann-Pick C1-like protein 1 (NPC1L1) and some are re-secreted by ABCG5 and ABCG8. In intestinal cells, cholesterol is packaged with triacylglycerols into chylomicrons. In hepatocytes, cholesterol is recycled or synthesized *de novo* by a pathway in which HMGCR is rate limiting. LDLs transport cholesterol from the liver to the periphery, where they are endocytosed. In HDL-cholesterol metabolism, APOA1 in HDLs mediates reverse cholesterol transport by interacting with ABCA1 and ABCG1 transporters on non-hepatic cells. LCAT esterifies cholesterol for the use in HDL-cholesterol, which is remodeled by CETP and by endothelial lipase (LIPG), in order to enter hepatocytes

expected to show efficient cholesterol lowering effects in patients. The inverse correlation between HDL levels (causing increased triacylglycerol levels) and the risk of CHD is due to the importance of HDL-cholesterol in reverse cholesterol transport from the periphery to the liver (Sect. 7.3). This initiated the search for HDL raising compounds, such as CEPT inhibitors, but they are far less efficient in preventing CVD than statins. The association between HDL-cholesterol and CHD is more complex than initially assumed, since HDLs contain a variety of proteins that are implicated in a number of biological pathways, such as oxidation, inflammation, hemostasis and innate immunity. This heterogeneity in the biological function of HDLs suggests that the measurement of HDL-cholesterol levels alone is insufficient for reflecting the role of HDLs in atherosclerosis.

Table 11.2 Phenomic and molecular description of selected dyslipidemias. Dyslipidemia names and their defining features are listed in rows and columns, respectively. The color intensity indicates for biochemical traits (*white*: no difference from normal; *red*: a fold-increase above normal; *blue*: a fold-decrease below normal), for susceptibility to CHD, stroke and peripheral vascular disease (*white*: no difference from normal; *red*: a fold-increase risk above normal; *blue*: a fold-decrease risk below normal), for qualitative clinical features (*white*: absence of feature, i.e. normal state; *red*: presence of the feature), for rare mutations (*white*: no role; *red*: a major etiologic role for the gene; *pink*: a minor etiologic role) and for common polymorphisms (*white*: no role; gradations of red: risk associated with the genotype)

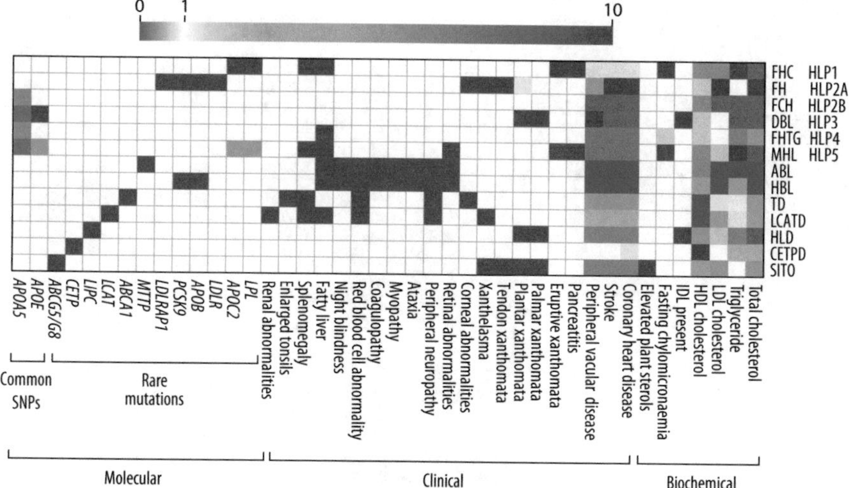

11.4 Dyslipidemias

Levels of certain plasma lipids and lipoproteins are key risk factors for CVD. Some 10 % of the hypercholesterolemia cases have a monogenic basis, such as heterozygous familial hypercholesterolemia, familial defective APOB and autosomal dominant hypercholesterolemia that is based on a gain of function of the *PCSK9* gene (Table 11.2). In contrast, different homozygous loss-of-function mutations in the genes *APOB* or *PCSK9* cause homozygous hypobetalipoproteinemia (HBL), in which almost no LDL-cholesterol is present. Homozygous mutations in *MTTP* cause a similar disease called abetalipoproteinemia (ABL). Rare monogenic disorders, such as Tangier disease (TD), affect HDL levels and are based on homozygous mutations in the *ABCA1* gene or deficiencies in the genes *APOA1*, *LCAT*, *CETP* or *LIPC*. Moreover, there are also rare monogenic disorders causing severe hypertriglycerolemia (HTG) that are due to homozygous loss-of-function mutations of the genes *LPL*, *APOC2* and *APOA5*. Very low triacylglycerol levels are found in patients with both ABL and HBL. Like in the case of monogenetic forms of obesity (Sect. 8.5) and T2D (Sect. 10.4), the identification of the genes causing monogenic dyslipoproteinemias significantly increased the understanding of the disease. For

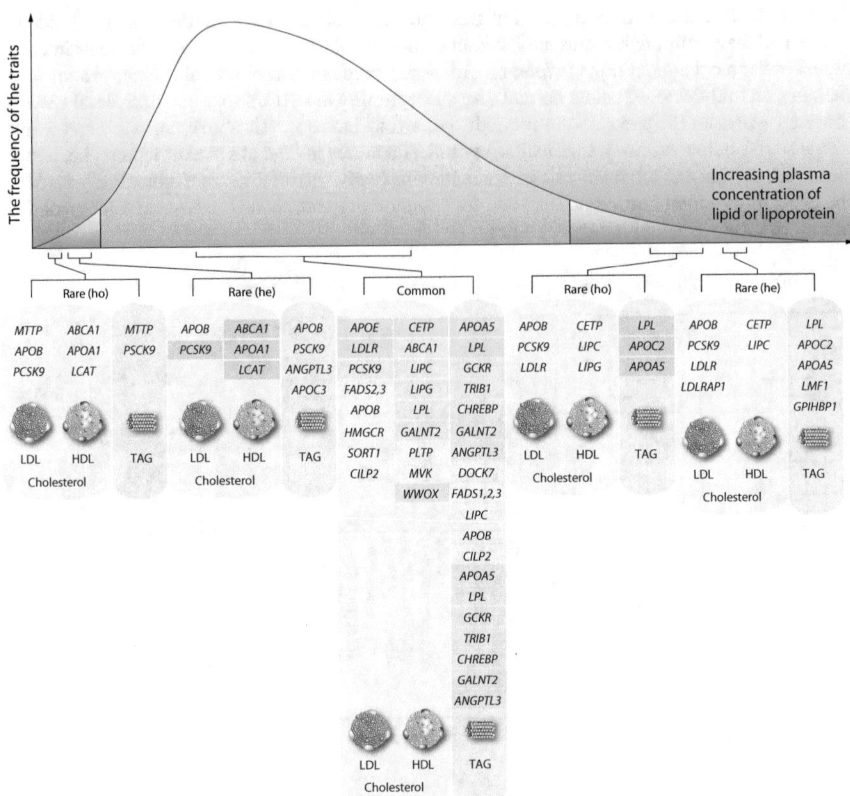

Fig. 11.6 Genetic variants affecting plasma lipoprotein distribution. The frequency of the traits LDL-cholesterol, HDL-cholesterol and triacylglycerol levels (y axis) is plotted over plasma concentrations (x axis). The bottom and top fifth percentiles of the distribution are indicated by shaded areas. The genes (*no shading*: by classical genetic or biochemical methods; *orange*: by re-sequencing; *blue*: by GWAS) that determine lipoprotein concentrations in specific segments of the distribution are shown below the respective graphs. The extremes of the distribution represent homozygotic (ho) monogenic disorders, the less extremes heterozygous (he) mutations and the center common variants. The green shading indicates small to moderate effect sizes associated with severe HTG

example, plasma LDL-cholesterol levels depend crucially on LDLR function, which in turn requires proper binding of APOB, the presence of LDLR accessory protein 1 (LDLRAP1) and intracellular LDLR degradation by PCSK9.

The majority of hypercholesterolemia cases is based on common variants in genes, such as *APOE*, *LDLR*, *APOB*, *PCSK9* and *HMGCR* for LDL-cholesterol and *CETP*, *LIPC*, *LPL*, *ABCA1*, endothelial lipase (*LIPG*) and *LCAT* for HDL-cholesterol, that display low ORs or epigenetic programming (Sect. 5.5). The genetic variants were identified either by SNP arrays or by targeted re-sequencing (Fig. 11.6). For example, an excess of mutations in the *NPC1L1* gene causes low intestinal sterol absorption or heterozygous mutations in the *LIPG* gene lead to high HDL levels.

The variations in lipoprotein levels that are based on common genetic variants are often too small to be meaningful in the clinical practice. However, these insights have a high value in basic research for the identification of novel pathways. Moreover, (epi)genetic profiling of human individuals for metabolic diseases, such as dyslipidemias, allows a genetic risk stratification far earlier than the onset of the metabolic syndrome (Chap. 12). Such personalized medicine approaches will provide for the respective human individuals longer and more effective periods of lifestyle changes. Since diet is a key determinant of lipoprotein levels, early dietary interventions are the most efficient and most economic strategies for CVD prevention. Furthermore, the results of biochemical assays, such as measurements of lipoprotein serum levels, should be integrated from multiple time points of the patient's lifetime.

Integrative genomic approaches can combine large-scale genome- and transcriptome-wide data, in order construct gene networks that underlie metabolic traits, such as plasma lipoproteins levels. For example, plasma HDL-cholesterol levels were linked to variants in the regulatory region of the vanin 1 (*VNN1*) gene using RNA expression profiles in lymphocytes. This type of analysis indicates that metabolic traits are the products of molecular networks being modulated by sets of complex genetic loci and environmental factors. This re-emphasizes that the genetic predisposition for a metabolic disease, such as atherosclerosis, is comprised of multiple common genetic variants that each have a small to moderate effect on the trait, either alone or in combination with modifier genes or environmental factors.

Future View

Atherosclerosis is the most important disease linked to chronic inflammation. Thus, a better understanding of the regulation of macrophage polarization will provide insights into pathways that could be used for the potential manipulation of macrophage behavior towards an athero-protective state. Next-generation whole genome sequencing will provide an unbiased approach for the identification of additional causative genes in patients with extreme lipoprotein phenotypes. This information needs to be integrated with metabolic states, such as obesity and T2D. Since diet plays a key part in the management of dyslipidemias, rational nutrigenetic studies should be undertaken. Defining new pathways and targets will allow new drug design and eventually lead to evidence-based changes in clinical practice. A prediction of the evolution and consequences of dyslipidemias in human individuals has to take into account the confounding influence of the environment, non-linear interactions between genes and environment and stochastic effects in the underlying genetic network.

Key Concepts

- Hypertension is the most important preventable risk factor for pre-mature death worldwide. It is defined as the blood pressure level above which therapeutic intervention has clinical benefit.
- Chronic hypertension in combination with atherosclerosis is the major risk factor for stroke, CHD, congestive heart failure and end-stage renal disease.

- Dietary factors, such a salt intake, significantly influence blood pressure, and reduced dietary salt intake as well as increased consumption of fruits and low fat food, exercise, weight loss and reduced alcohol intake can reduce hypertension.
- Atherosclerosis is a chronic disease of large and medium arteries, in which cholesterol deposition below the endothelium causes a macrophage-dominated inflammatory response.
- Hypercholesterolemia is important for the recruitment of macrophages into the arterial wall, but also immunological and mechanical injuries, as well as bacterial and viral infections, contribute to the pathogenesis of atherosclerosis.
- Macrophages internalize native LDLs and VLDLs as well as oxidized lipoproteins in the plaque via macropinocytosis, phagocytosis of aggregated LDLs and scavenger receptor-mediated uptake.
- Dys-regulation of lipid metabolism in foam cells contributes to ER stress ultimately resulting in apoptotic cell death.
- Originally stable lesions change into unstable vulnerable plaques that can easily lead to the rupture of the endothelium, leading to attachment of cell debris and/or blood clots that can cause myocardial infarction or cerebral stroke.
- In contrast to most other chronic inflammatory diseases, in atherosclerosis there is the potential to remove the inflammatory stimulus.
- Only a small amount of circulating cholesterol originates from the diet, while approximately 80 % is derived from endogenous synthesis.
- There are four main types of lipoproteins: chylomicrons, VLDLs, LDLs and HDLs, which are classified based on their density and size.
- Some 10 % of the hypercholesterolemia cases have a monogenic basis. Like in case of monogenetic forms of obesity and T2D, the identification of the genes causing monogenic dyslipoproteinemias significantly increased the understanding of the disease.
- (Epi)genetic profiling of human individuals for metabolic diseases, such as dyslipidemias, allows a genetic risk stratification far earlier than the onset of the metabolic syndrome.
- Integrative genomic approaches can combine large-scale genome- and transcriptome-wide data, in order construct gene networks that underlie metabolic traits, such as plasma lipoproteins levels.

Additional Reading

Hegele RA (2009) Plasma lipoproteins: genetic influences and clinical implications. Nat Rev Genet 10:109–121

Jensen MK, Bertoia ML, Cahill LE, Agarwal I, Rimm EB, Mukamal KJ (2014) Novel metabolic biomarkers of cardiovascular disease. Nat Rev Endocrinol 10:659–672

Moore KJ, Sheedy FJ, Fisher EA (2013) Macrophages in atherosclerosis: a dynamic balance. Nat Rev Immunol 13:709–721

Swirski FK, Nahrendorf M (2013) Leukocyte behavior in atherosclerosis, myocardial infarction, and heart failure. Science 339:161–166

Chapter 12
The Metabolic Syndrome

Abstract Increased consumption of high-caloric diets combined with reduced physical activity are the main causes for the worldwide dramatic increase in the metabolic syndrome. Obesity (Chap. 8) and insulin resistance (Chap. 9) is initiating the development of the metabolic syndrome, and both significantly increase the risk of T2D (Chap. 10) and CVD (Chap. 11). Insulin plays a central role in regulating energy homeostasis in metabolic tissues. The effects of hyperglycemia and insulin resistance on the risk of CVD are largely pathway- and tissue-specific. The genetic risk for the metabolic syndrome overlaps with that of its major components, such as obesity, T2D and dyslipdemia, but like in these traits, common genetic variations can explain only a minor part of the disease risk. However, there is emerging evidence for an important role of epigenetics, causing heritable changes in gene expression, in the origin and development of the metabolic syndrome.

In this chapter, we discuss the role of insulin resistance and obesity in the major metabolic tissues liver, skeletal muscle, pancreas and WAT causing the metabolic syndrome. We will emphasize pathways that are involved in energy metabolism and will describe how insulin resistance and obesity lead to the metabolic syndrome and its complications. The importance of inflammation and regulation of energy metabolism will be highlighted. In addtion, we will discuss the future challenges and possibilities to treat and prevent the metabolic syndrome by dietary modifications.

Keywords Obesity • Insulin resistance • T2D • Metabolic syndrome • Liver • Adipose tissue • Skeletal muscle • Pancreas • Macrophages • Inflammation • Epigenetics

12.1 Definitions of the Metabolic Syndrome

The metabolic syndrome is nowadays a very common aging-related condition that occurs primarily as a result of overweight and obesity caused by a sedentary lifestyle, i.e. physical inactivity, and the consumption of diet containing excess of calories (Sect. 8.3). The syndrome is composed of different factors that either alone or in combination significantly increase the risk of T2D and CVD (Chaps. 10 and 11). Most of these risk factors have been discussed in the previous chapters, such as visceral obesity (Sect. 8.1), ectopic lipid overload (Sect. 9.4), insulin resistance

C. Carlberg et al., *Nutrigenomics*, DOI 10.1007/978-3-319-30415-1_12

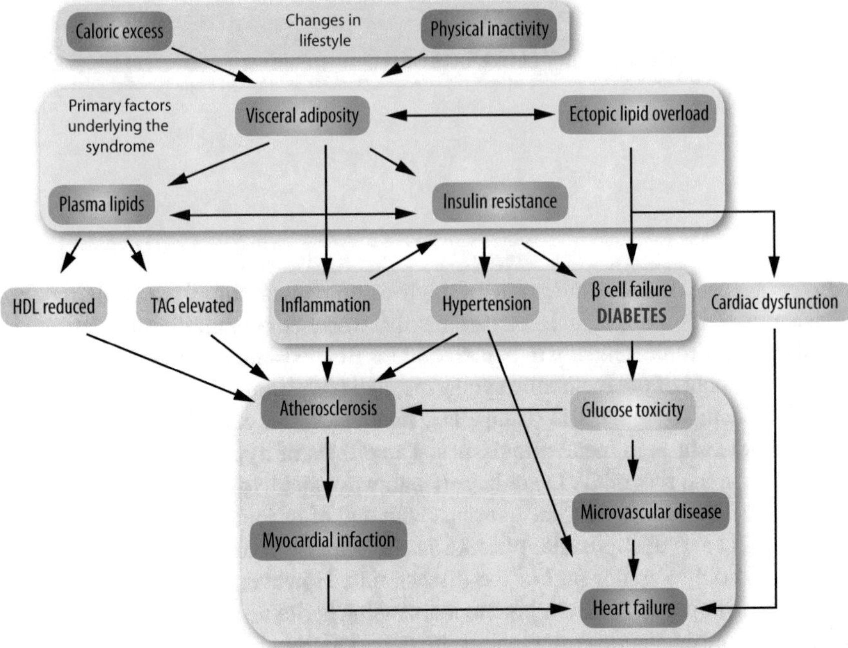

Fig. 12.1 Interactions of metabolic syndrome traits in T2D and CVD. Changes in lifestyle, such as increased consumption of high-caloric diets combined with reduced physical activity play important roles in the worldwide dramatic increase in the metabolic syndrome. Visceral obesity, ectopic lipid overload, dyslipidemias and insulin resistance are the primary factors underlying the syndrome. These factors cause inflammation, hypertension and β cell failure. Individuals with the metabolic syndrome are therefore at increased risk for the development of atherosclerosis, T2D, microvascular diseases, myocardial infarction and finally heart failure

(Sect. 9.4), β cell failure (Sect. 9.5), hypertension (Sect. 11.1) and dyslipidemias with high plasma concentrations of triacylglycerols and low concentrations of HDL-cholesterol (Sect. 11.4) (Fig. 12.1). The dramatic worldwide increase in obesity (Sect. 8.1) and the parallel rise in life expectancy, i.e. the increased number of elderly around the world, make this condition a major global health problem.

Historically, the concept of "syndrome X" was used to describe the metabolic syndrome in the end of 1980s as a condition with increased risk of T2D and CVD caused by insulin resistance in metabolic tissues. More recent, the *National Cholesterol Education Program* (*NCEP*), the *WHO*, the *European Group for the study of Insulin Resistance* (*EGIR*) and the *International Diabetes Federation* (*IDF*) used slightly different thresholds to define the metabolic syndrome based on rate of obesity, hyperglycemia, dyslipidemias and hypertension (Table 12.1). Moreover, while the *NCEP* definition does not require any defined parameter, the *WHO* proposes that an evidence of insulin resistance, such as impaired glucose tolerance (IGT), impaired fasting glucose (IFG) or T2D, is essential. In contrast, the *EGIR*

Table 12.1 Definitions of the metabolic syndrome

	NCEP (2005)	WHO (1998)	EGIR (1999)	IDF (2005)
Absolutely required	None	Insulin resistance	Hyperinsulinemia	Central obesity
Criteria	Any of the 5 below	Insulin resistance or T2D	Hyperinsulinemia	Obesity
		Plus 2 of 5 below	Plus 2 of the 4 below	Plus 2 of the 4 below
Obesity	Waist circumference:	Waist/hip ratio:	Waist circumference:	Central obesity
	Males > 101.6 cm,	Males > 0.90,	Males > 94 cm,	
	Females > 88.9 cm	Females > 0.85 or BMI > 30 kg/cm^2	Females > 80 cm	
Hyperglycemia	Fasting glucose > 5.6 mM	Insulin resistance	Insulin resistance	Fasting glucose > 5.6 mM
Dyslepidemia I	Triacylglycerols > 1.7 mM	Triacylglycerols > 1.7 mM	Triacylglycerols > 2.0 mM	Triacylglycerols > 1.7 mM
		or HDL < 0.9 mM	or HDL < 1.0 mM	
Dyslepidemia II	HDL:			HDL:
	Males < 1.0 mM,			Males < 1.0 mM,
	females < 1.25 mM			females < 1.25 mM
Hypertension	>130 mm Hg systolic or	>140/90 mm Hg	>140/90 mm Hg	>130 mm Hg systolic or
	>85 mm Hg diastolic			>85 mm Hg diastolic
Other criteria		Microalbuminuria		

emphasizes hyperinsulinemia as main criterium, while for the IDF central obesity is essential. At present, the definitions of *NCEP* and *IDF* are most widely used.

12.2 Whole Body's Perspective on the Metabolic Syndrome

The human body has developed integrated mechanisms to become either catabolic, when energy demands cannot be met by food intake, or anabolic, when calorie supply exceeds energy demands. The key regulator of these mechanisms is insulin (Chaps. 9 and 10), which is secreted by β cells in the pancreas after a meal and promotes carbohydrate resorption, energy utilization (via glycolysis), storage of carbohydrates as glycogen, storage of fat (as triacylglycerols), synthesis of fat from carbohydrates (via activating *de novo* lipogenesis) in key metabolic tissues. At the same time, insulin inhibits lipolysis, i.e. the release of energy from triacylglycerols, and the synthesis of glucose (via gluconeogenesis) after a meal. Thus, the actions of insulin create an integrated set of signals that represent the nutrient availability and the energy demands of the human body. In turn, a disturbance in insulin actions, such as obesity-triggered insulin resistance in one or multiple metabolic organs, such as skeletal muscle, liver and WAT, often serves as the onset of the metabolic syndrome (Fig. 12.2). These conditions can lead to organ-specific consequences, such as β cell failure and NAFLD, but also to systemic effects, such as glucotoxicity, lipotoxicity and low-grade inflammation. All these conditions are key factors of the metabolic syndrome and accelerate the risk of diabetes, heart disease and their complications.

12.3 Metabolic Syndrome in Key Metabolic Organs

Systemic effects of the metabolic syndrome influence the metabolism in key metabolic organs, such as liver, muscle, pancreas and WAT. Hepatic insulin resistance causes elevated activity of the key gluconeogenesis enzymes G6PC and PCK, and increased glycogenolysis, both leading to increased glucose output from the liver (Fig. 12.3). In parallel, the expression of enzymes that regulate glycogen synthesis and glycolysis, such as GCK and pyruvate kinase, is reduced, driving GLUT2 to transport glucose out of hepatocytes. All these alterations in the glucose metabolism pathways accelerate systemic glucotoxicity. In addition, a decreased insulin sensitivity in the liver increases the uptake of FFAs and the formation of triacylglycerols. These are loaded to VLDLs for transport in the circulation and thus cause dyslipidemia. The increase of glucose levels in the liver increases lipogenesis via the activity of SREBF1 and FASN, which are both not impaired by insulin resistance. This accumulation of lipids in liver can cause NAFLD. Hepatic insulin resistance also contributes to hyperlipidemia through the down-regulation of LDLR (Sect. 11.4). In

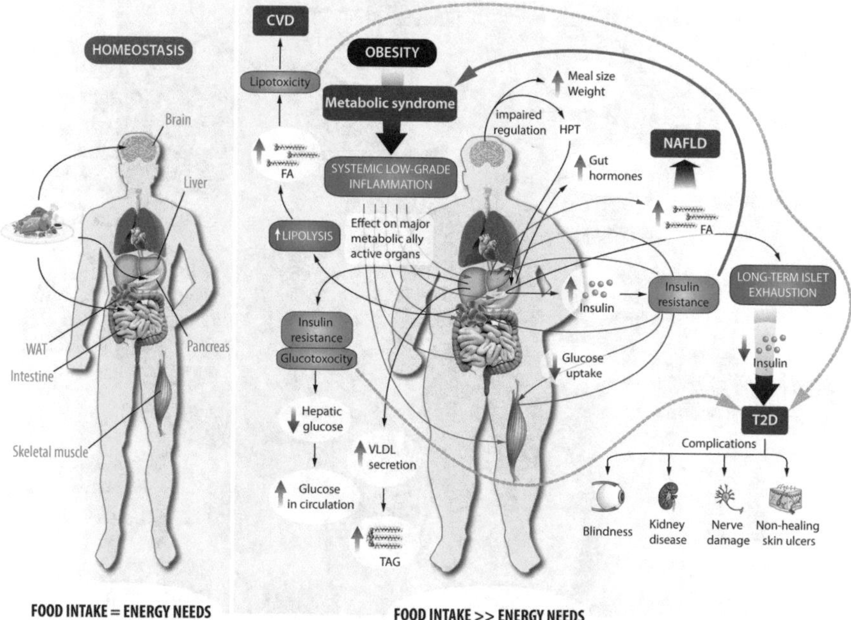

Fig. 12.2 Whole body's view on the metabolic syndrome. Under normal conditions, energy intake and utilization is perfectly balanced with the body's energy needs. Intestine, pancreas and brain sense food after a meal and send signals to muscle, liver, fat and back to the brain, in order to maintain metabolic homeostasis via the coordination of uptake and storage of nutrients and energy production. The metabolic syndrome often starts with obesity, triggering a state of systemic low-grade inflammation that affects various major organs involved in metabolic homeostasis (Chaps. 7 and 8). The brain's ability to regulate meal size or frequency is impaired leading to weight gain and further organ dysfunction. The autonomic nervous system and the hypothalamic-pituitary-thyroid (HPT) axis are disrupted causing a change in the release of gut hormones (Sect. 8.4). Insulin resistance is a further important trigger of the metabolic syndrome (Sect. 9.4). In the pancreas, the islets expand to release more insulin (hyperinsulinemia) in an attempt to overcome insulin resistance of muscle, liver and WAT. However, over time the islets become exhausted and little or no insulin is produced, so that T2D occurs (Chap. 10). Insulin resistance in the muscle leads to excessive glucose uptake in the liver that is primarily converted to fatty acids often causing NAFLD. Moreover, in the liver, glucotoxicity and insulin resistance result in inefficient down-regulation of hepatic glucose production leading to further increase of circulating glucose levels. The fat excess in the liver can be released into the circulation as VLDLs leading to elevated serum triacylglycerol levels. Insulin resistance of adipose tissue increases its lipolytic activity, thus also releasing excess fatty acids. All together these lipid sources result in lipotoxicity that further contributes to organ dysfunction and disease, especially CVD. Lipotoxicity and glucotoxicity worsen T2D and lead to numerous complications, such as kidney disease, blindness, nerve damage and non-healing skin ulcers

this way, liver insulin resistance causes decreased clearance of LDLs and VLDLs, leading to increased LDL and VLDL levels in the circulation, respectively.

Skeletal muscle is the main tissue for glucose storage and utilization, and approximately 80 % of the glucose load of the blood stream after a meal are taken up by the muscle. Insulin resistance in muscles leads to reduced insulin-stimulated

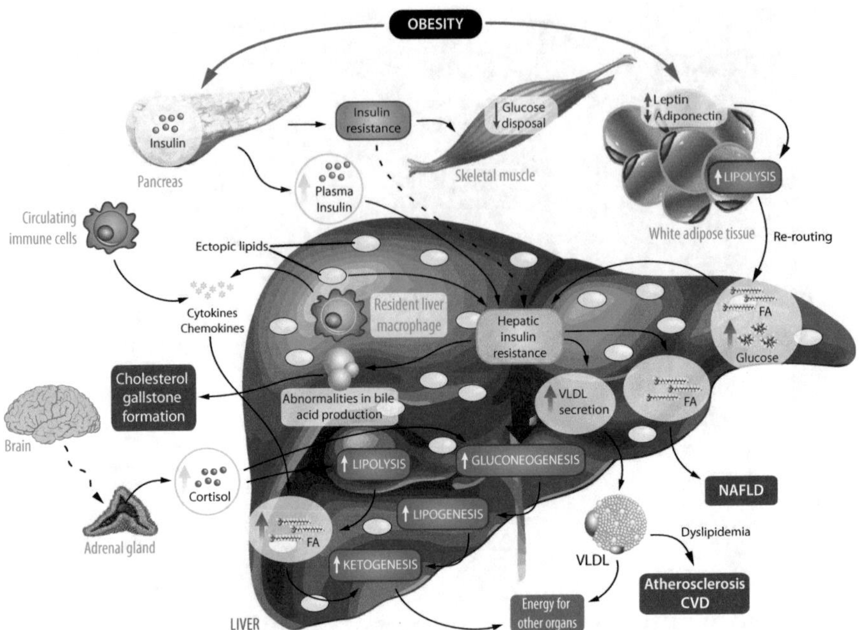

Fig. 12.3 Metabolic syndrome in the liver. Early stages of the metabolic syndrome are associated with insulin resistance causing a decreased glucose disposal in skeletal muscle. Obesity also alters the secretion pattern of adipokines, such as increased levels of leptin and decreased adiponectin concentrations (Sect. 8.6). Together this leads to a re-routing of glucose and lipids (as a consequence of increased lipolysis from adipose tissue) to the liver. As intracellular lipid levels in the liver rise, along with the increased plasma insulin, hepatic insulin signaling rapidly deteriorates and this impairment turns the liver into a "co-conspirator" in the further progression of the metabolic syndrome and its complications. Hepatic insulin resistance leads to increased hepatic glucose production. Further hormones can contribute to decreased insulin sensitivity of the liver. For example, cortisol increases glucose production, promotes lipolysis and increases lipid deposition. Insulin resistance increases the secretion of VLDLs from the liver and causes dyslipidemia. Excessive hepatic secretion of VLDLs play an important role in promoting atherosclerosis and CVD (Sect. 11.3). Insulin resistance also increases intra-hepatic fat accumulation finally causing to NAFLD. Activation of inflammatory pathways occurs both systemically, as a result of cytokines released from circulating immune cells, and locally through resident liver macrophages. This further accelerates lipid accumulation and storage. Liver insulin resistance also causes abnormalities in bile acid production and increases the risk for cholesterol gallstone formation

GLUT4-mediated glucose transport into the myocytes (Fig. 12.4). Decreased glucose uptake reduces the levels of glucose-6-phosphate to be used for glycogen synthesis and glycolysis. This increases the glucose concentration in the bloodstream and causes systemic glucotoxicity. Like the liver, systemic lipotoxicity also overloads the muscle with FFAs. These lipids are taken up and stored in form of triacylglycerols in intra-muscular lipid droplets.

In the early stages of insulin resistance, β cells increase the production and secretion of insulin, in order to maintain glucose tolerance (Fig. 12.5). Since insulin under these conditions is less potent in suppressing hepatic glucose production, the

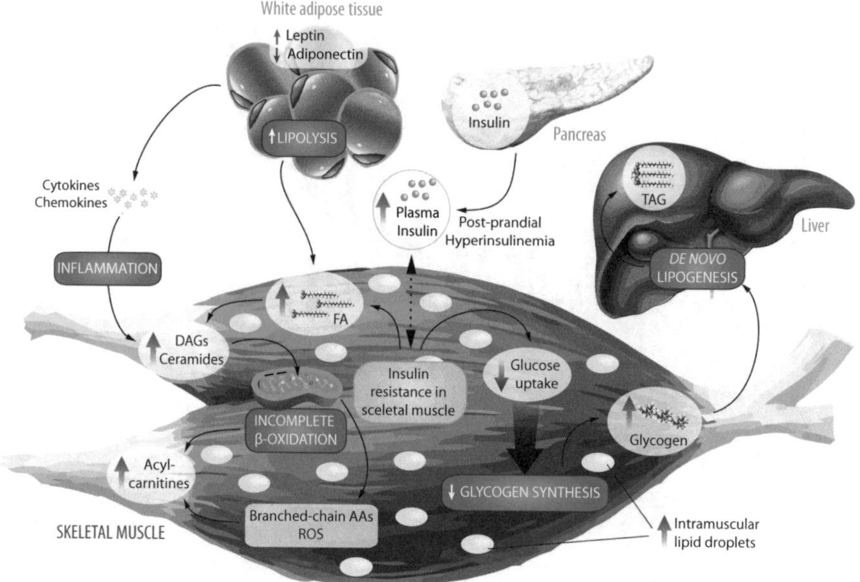

Fig. 12.4 Metabolic syndrome in skeletal muscles. Insulin resistance in skeletal muscle causes reduced insulin-stimulated glucose uptake and therefore less glucose is available for insulin-stimulated glycogen synthesis. Overload of lipids increases the level of intra-myocellular lipids in the form of DAG and ceramides as well as increases in acyl-carnitines (due to incomplete mitochondrial fatty acid oxidation). Pro-inflammatory adipokines, branched-chain amino acids and ROS further contribute to this defect in insulin signaling. Insulin resistance in skeletal muscle promotes post-prandial hyperinsulinemia and diversion of ingested carbohydrates away from storage as glycogen in skeletal muscle to liver where they are converted to triacylglycerols through increased hepatic *de novo* lipogenesis

liver becomes insulin resistant. When insulin resistance progresses, β cells lose their ability to compensate for decrease insulin response via an increase of insulin release. This finally results in reduced circulating insulin concentrations and often comes along with increased glucagon levels. This shift in the glucagon-insulin ratio leads to a further rise in hepatic gluconeogenesis and advanced hyperglycemia occurs. Systemic glucotoxicity and lipotoxicity, i.e. constant exposure of β cells to elevated levels of glucose and lipids, both increase glucose metabolism in β cells and cause metabolic stress leading to the unfolding protein response of the ER in these cells. In response to ER stress, hypoxic stress and pro-inflammatory cytokines, the β cells fail to proliferate and undergo uncontrolled autophagy or even apoptosis. This leads to β cell dysfunction and ultimately their death.

In obesity, the storage capacity of adipocytes is often exceeded causing cellular dysfunctions, such as increased formation of ceramides, ER stress and hypoxia leading to reduced metabolic control and cell death. An increase in number and size of adipocytes also influences the secretion of adipokines. In addition, adipocytes attract monocytes into WAT that become M1-type macrophages, secreting pro-inflammatory cytokines, together with adipokines leading to low-grade sys-

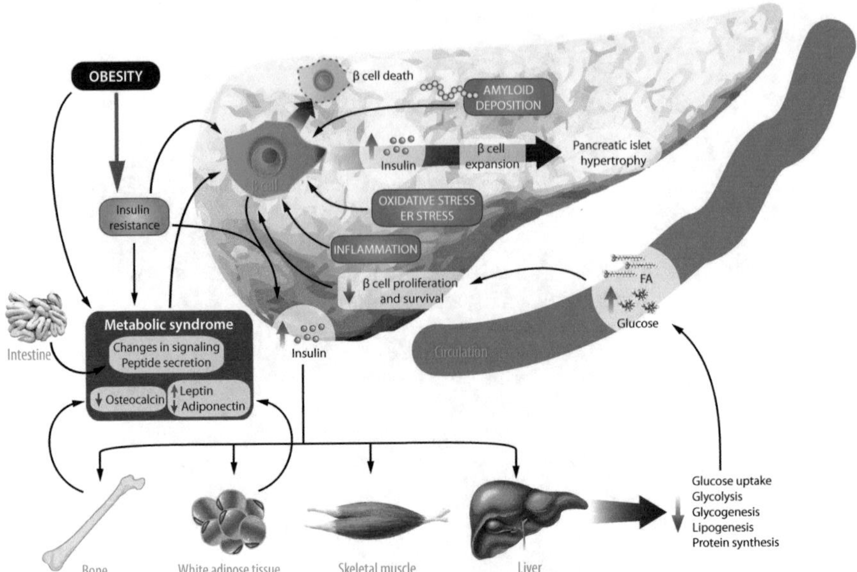

Fig. 12.5 Metabolic syndrome in the pancreas. Obesity promotes the development of insulin resistance leading to compensatory increases in insulin release from β cells. Chronic overproduction of insulin leads to β cell expansion, i.e. pancreatic islet hypertrophy. As insulin resistance progresses, the effects of insulin on target tissues diminish, thereby leading to impairments in glucose uptake, glycolysis, glycogenesis, lipogenesis and protein synthesis. This leads to hyperglycemia and elevated FFA levels in the circulation that negatively affect β cell proliferation and survival. This results in a vicious cycle that further reduces β cell function. In the setting of the metabolic syndrome, many tissues show alterations in hormone levels that directly impact β cell function. The intestine changes the secretion of signaling peptides, the bone secretes less osteocalcin and WAT secretes more leptin but less adiponectin. These hormonal changes together with oxidative stress, ER stress, inflammation and intracellular amyloid deposition cause β cell death (Sect. 9.5)

temic inflammation. All this contributes to insulin resistance in WAT causing a reduced insulin-stimulated import of glucose via GLUT4 (Fig. 12.6). Moreover, lipolysis is increased due to the impaired inhibition of LIPE activity leading to an increase in FFA release from adipocytes. The ability of insulin to stimulate the re-esterification of FFAs is also impaired, and consecutively systemic lipotoxicity occurs.

The metabolic and inflammatory response of metabolic tissues, such as WAT, integrates the actions of the innate immune system, i.e. primarily of the monocyte-derived macrophages, with those of adipocytes (Fig. 12.7). This integrated action of two or more different tissues has developed during evolution. Responses of the immune system to pathogen invasions use substantial amounts of energy for new protein synthesis and rapid growth of cells. Thus, it makes sense that metabolic organs are insulin insensitive, i.e. that they use less energy from circulating glucose and lipids during limited time periods. For this reason, inflammatory mediators control energy metabolism so that pathogens are defended most efficiently, for example through the ability to shift rapidly from glucose oxidation to lipid oxidation. With a similar protective attempt lipids can trigger insulin resistance, in order

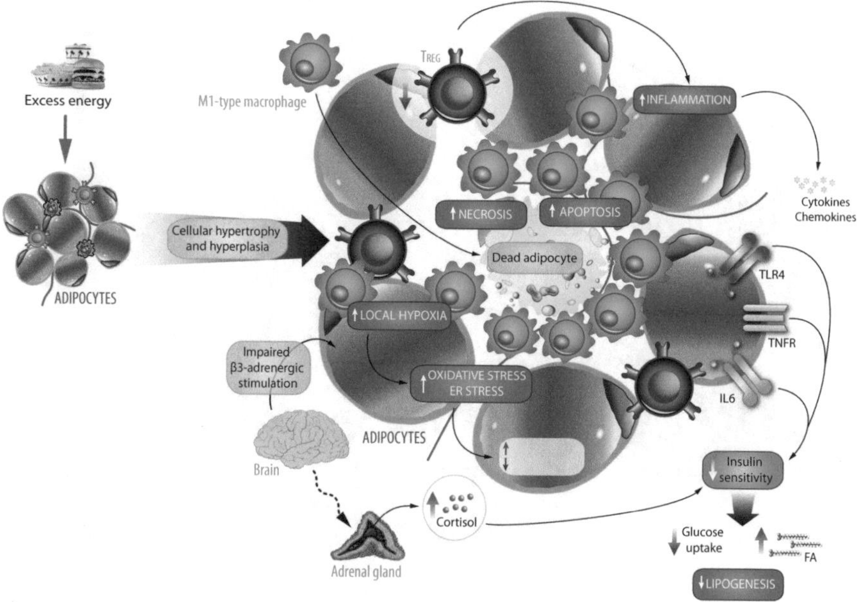

Fig. 12.6 Metabolic syndrome in WAT. In the presence of excess energy supply WAT expands as a result of cellular hypertrophy and hyperplasia. These enlarged adipocytes become dysregulated through an increased rate of local necrosis, apoptosis and pro-inflammatory responses. Dead adipocytes attract macrophages that are conventionally skewed towards an M1-like pro-inflammatory profile. Under obese conditions there is also a reduction of T_{REG}. This causes an increase in the local pro-inflammatory environment that can ultimately spill over to systemic increases in pro-inflammatory cytokines. The rapid tissue expansion during obesity leads to local hypoxia and the activation of ER stress response causing a reduced release of insulin-sensitizing adipokines, such as adiponectin (Sect. 8.6). Moreover, increased levels of cortisol and the activation of TLRs and other pro-inflammatory cytokine receptors, such as TNFR and IL6 receptors, further reduce insulin sensitivity. This leads to reduced rates of triacylglycerol synthesis, increasing levels of FFAs and a decrease in insulin-mediated glucose uptake. In contrast, the impaired β3-adrenergic response downstream of SNS activity leads to reduced metabolic flexibility, since FFAs cannot be appropriately activated in response to β3-adrenergic stimulation

to preserve glucose for glucose-dependent organs, such as the CNS and erythrocytes.

12.4 Genetic and Epigenetic Basis of the Metabolic Syndrome

GWAS analysis for central factors of the metabolic syndrome, such as BMI, T2D and dyslipidemia, have identified for each of these traits highly statistically significant associations with 40 to 100 genetic variants (Sects. 8.4, 10.4 and 11.4). The key genes of these lists, such as *LPL*, *APOE*, *MC4R*, *FTO* and *TCF7L2* (Table 12.2), are also the central determinants for the genetic risk for the metabolic syndrome.

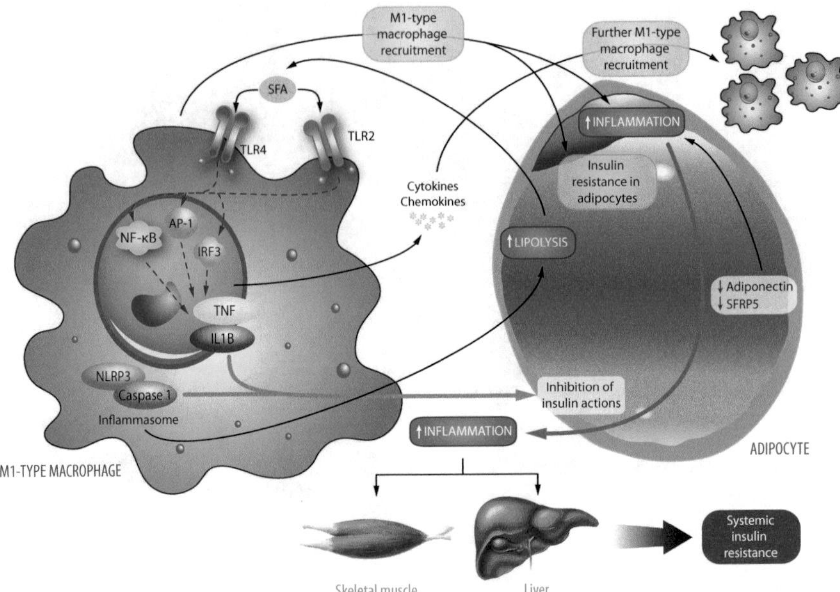

Fig. 12.7 Metabolic syndrome driven by the interaction of macrophages with WAT. Obesity recruits M1-type macrophages into adipose tissue that promote local inflammation and insulin resistance. SFAs stimulate TLR2 and TLR4 in these M1 macrophages leading to the activation of the pro-inflammatory transcription factors IRF3, AP-1 and NF-κB. This induces the production of inflammatory cytokines, such as IL1B and TNF, that inhibit insulin actions in neighboring adipocytes. The production of IL1B is also augmented by activation of the NLRP3 inflammasome inducing lipolysis. In a feed-forward loop, the released fatty acids from the increased lipolysis induce the expression of chemokines leading to the recruitment of further M1-type macrophages into adipose tissue. Reduced circulating levels of the anti-inflammatory adipokines adiponectin and SFRP5 potentiate inflammation. Finally, as inflammation spreads from adipocytes to other organs, such as skeletal muscle and liver, systemic insulin resistance ensues

Table 12.2 Central genes in the development of metabolic syndrome

Gene	Gene function	Disease context	Affected parameter
LPL	Hydrolyzes triacylglycerols (Sect. 3.4)	Cardiovascular	HDL concentration
APOE	Removing lipoproteins from circulation (Sect. 11.4)	Cardiovascular	HDL concentration
MC4R	Membrane receptor on neurons binding α-MSH (Sect. 8.4)	Obesity	Waist circumference
FTO	Exact function unknown (Sect. 8.7)	Obesity	Waist circumference
TCF7L2	Transcription factor in β cells (Sect. 10.3)	T2D	Glucose concentration

However, the dilemma with all these common SNPs remains that their individual ORs are clearly below 2, i.e. they contribute considerably less than 100 % increased risk for the disease. This implies that the common SNPs can explain only a small fraction of the cases of the metabolic syndrome.

Within slightly more than one generation (33 years) the worldwide prevalence for obesity doubled (Sect. 8.1). Very obviously the Western-style high-fat diet combined with decreased physical activity are the main environmental contributors to obesity and the subsequent development of the metabolic syndrome. In addition, human populations that made a transition from famine to food surplus just within 1–2 generations are under significantly higher risk for obesity, T2D and the metabolic syndrome, than those that were improving their nutritional conditions over many generations. This means that people who are born and live in countries that had particularly rapid changes in urbanization and economic development have an increased risk of the metabolic syndrome in the years to come. This suggests that both socio-environmental factors and epigenetic mechanisms rather than variations of the core genome play a role in the obesity epidemic and its associated metabolic abnormalities.

There are more and more epidemiological and clinical evidences that the concept of the thrifty hypothesis (Sects. 5.5 and 10.5), i.e. a pre-natal epigenetic programming *in utero*, may be the key cause for metabolic diseases. Persons that carry an epigenome that during their anthropologic development was programmed by sub-optimal nutrition *in utero*, can despite normal post-natal nutrition transgenerationally transmit a predisposition for obesity (Fig. 12.8). So far, there is no comprehensive analysis of the epigenome of persons suffering from the metabolic syndrome, i.e. no concrete genomic regions of elevated risks have been identified. However, it can be assumed that due to the complexity of insulin signaling and its interference with multiple pathways (Sect. 9.2) a high number of regions will be affected in a individual-specific way. Nevertheless, since epigenetic modifications respond dynamically to environmental conditions (Chap. 5), there is potential for intervention and reversibility.

Future View

Healthy dietary patterns, such as Mediterranean or Nordic diet, lower the risk of the metabolic syndrome. In future, studies understanding the molecular mechanism of diet on epigenetic programming in the pre-natal, post-natal and adult phases of life will be of major importance to understand how diet can prevent the metabolic syndrome. Thus, cleverly designed dietary intervention studies and observational studies will investigate the impact of individual nutrients, such as vitamin D_3 or PUFAs, within a healthy dietary pattern, in order to improve the conditions of the metabolic syndrome. In these studies, a larger number of epigenome-, transcriptome-, proteome- and metabolome-wide data will need to be integrated. A futuristic way to measure these will be injectable nanosensors that perform a continuous surveillance of the blood. Such sensors will detect epigenome, RNA, protein and metabolite signals and may transmit them wirelessly to the individual's smartphone. For example, this allows an early warning of a heart attack via endothelial obliteration from

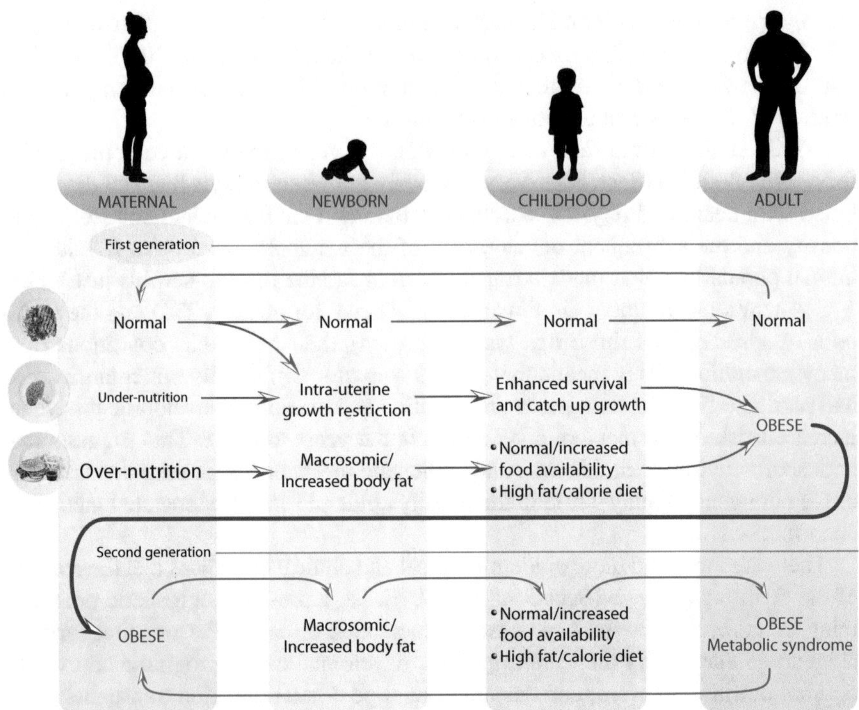

Fig. 12.8 Epigenetic programming and the shift of populations towards obesity and metabolic syndrome. Non-obese mothers usually give birth to non-obese children, which develop into adults with a normal metabolic profile and a normal body fat content. However, undernutrition combined with improved neo-natal survival, formula feeding and exposure to a Western post-natal diet increased the incidence of pre-maturity and intra-uterine growth restriction. This results in increased obesity of the offspring and higher risk for obtaining the metabolic syndrome. Some obese mothers may give birth to newborns with increased body fat, as a result of consumption of a high-fat diet. All these processes contribute to a shift of the population towards an obese phenotype. This also includes that second generation obese women have an increased risk to give birth to infants with increased body fat content and a further increased risk to develop obesity and the metabolic syndrome

an artery, the onset of T1D in a child via the detection of auto-antibodies against β cells before they get destroyed or the rise of fasting glucose levels after infectious diseases (Sect. 4.6). This surveillance concept has the potential to detect a risk signal in time and to implement better prevention and therapy by personalized nutrititon.

Key Concepts

- Increased consumption of high caloric diets combined with reduced physical activity play important roles in the worldwide dramatic increase in the metabolic syndrome.

- No universally accepted definition for the metabolic syndrome exists, but subjects with metabolic syndrome have a cluster of features including visceral obesity, hypertension, dyslipidemia and dysglycemia, increasing the risk of T2D and CVD.
- The dramatic worldwide increase of obesity and the parallel rise in life expectancy, leading to an increased number of elderly obese people, makes the metabolic syndrome a major global health problem.
- Excess food intake and β cell failure caused by insulin resistance and obesity leads to systemic glucotoxicity, systemic lipotoxicity and systemic low-grade inflammation. All these conditions are key components of the metabolic syndrome.
- All metabolic tissues are affected, and in the liver, the glucose output is increased due to increased glucose synthesis and increased breakdown of glycogen, caused by insulin resistance. At the same time, systemic lipotoxicity overloads liver with lipids that are taken up by the cells and form triacylglycerol, which are transported as VLDL in the circulation and causes dyslipidemia, but also accumulate to form fatty liver and NAFLD.
- In skeletal muscle, insulin resistance leads to reduced insulin-stimulated glucose uptake. Since skeletal muscle normally takes up 80 % of the glucose in the circulation, the reduced uptake, utilization and storage of glucose all increase the concentration of glucose in the bloodstream and cause systemic glucotoxicity. The skeletal muscle is also overloaded with lipids, which are stored in intramuscular lipid droplets.
- The β cells produce less insulin, which occurs in parallel with increased glucagon levels. The shift in the glucagon-insulin ratio leads to a further rise in hepatic synthesis. Systemic glucotoxicity and systemic lipotoxicity both increase glucose metabolism in β cells and cause ER stress and hypoxic stress, leading to β cell dysfunction and cell death.
- In WAT, lipolysis is increased, and uptake of FFAs is impaired, causing systemic lipotoxicity. Obesity increases the adipocytes number and size, which influences the secretion of adipokines.
- The metabolic and inflammatory response is integrated due to the role that these two pathways have played in evolution. The reason why inflammatory mediators impair insulin signaling is the cells ability to shift rapidly from glucose oxidation to lipid oxidation.
- The genes *LPL*, *APOE*, *MC4R*, *FTO* and *TCF7L2* are the central determinants of the genetic risk for the metabolic syndrome. However, common SNPs can explain only a small fraction of the cases of the metabolic syndrome.
- Human populations that made a transition from famine to food surplus just within 1–2 generations are under significantly higher risk for obesity, T2D and the metabolic syndrome than those that were improving nutritional conditions over many generations.
- There is no comprehensive analysis of the epigenome of persons suffering from the metabolic syndrome, but it can be assumed that due to the complexity of insulin signaling and its interference with multiple pathways that a high number of regions will be affected in a individual-specific way.

Additional Reading

Desai M, Jellyman JK, Ross MG (2015) Epigenomics, gestational programming and risk of metabolic syndrome. Int J Obes (Lond) 39(4)

Hotamisligil GS (2010) Endoplasmic reticulum stress and the inflammatory basis of metabolic disease. Cell 140:900–917

Huang PL (2009) A comprehensive definition for metabolic syndrome. Dis Model Mech 2:231–237

Keane D, Kelly S, Healy NP, McArdle MA, Holohan K, Roche HM (2013) Diet and metabolic syndrome: an overview. Curr Vasc Pharmacol 11:842–857

Lusis AJ, Attie AD, Reue K (2008) Metabolic syndrome: from epidemiology to systems biology. Nat Rev Genet 9:819–830

Nature Medicine: www.nature.com/nm/e-poster/eposter_full.html

Perez-Martinez P, Garcia-Rios A, Delgado-Lista J, Perez-Jimenez F, Lopez-Miranda J (2012) Metabolic syndrome: evidences for a personalized nutrition. Mol Nutr Food Res 56:67–76

Printed in the United States
By Bookmasters